Undergraduate Topics in Computer Science

For other titles published in this series, go to
www.springer.com/series/7592

Undergraduate Topics in Computer Science (UTiCS) delivers high-quality instructional content for undergraduates studying in all areas of computing and information science. From core foundational and theoretical material to final-year topics and applications, UTiCS books take a fresh, concise, and modern approach and are ideal for self-study or for a one- or two-semester course. The texts are all authored by established experts in their fields, reviewed by an international advisory board, and contain numerous examples and problems. Many include fully worked solutions.

Wilhelm Burger · Mark J. Burge

Principles of Digital Image Processing

Fundamental Techniques

Springer

Wilhelm Burger
University of Applied Science
Hagenberg, Austria
wilbur@ieee.org

Mark J. Burge
noblis.org
Washington, D.C.
mburge@acm.org

Undergraduate Topics in Computer Science ISSN 1863-7310
ISBN: 978-1-84800-190-9 e-ISBN: 978-1-84800-191-6
DOI: 10.1007/978-1-84800-191-6

British Library Cataloguing in Publication Data
A catalogue record for this book is available from the British Library
Library of Congress Control Number: 2008942779

springer.com

Burger & Burge: Principles of Digital Image Processing: Fundamental Techniques

First Indian Reprint 2012

ISBN 978-81-322-0496-1

This edition is published by Springer (India) Private Limited, (a part of Springer Science+Business Media), Registered Office: 7th Floor, Vijaya Building, 17 Barakhamba Road, New Delhi – 110 001, India.

Printed in India by Rakmo Press Pvt. Ltd., New Delhi.

Preface

This book provides a modern, algorithmic introduction to digital image processing, designed to be used both by learners desiring a firm foundation on which to build and practitioners in search of critical analysis and modern implementations of the most important techniques. This updated and enhanced paperback edition of our comprehensive textbook *Digital Image Processing: An Algorithmic Approach Using Java* packages the original material into a series of compact volumes, thereby supporting a flexible sequence of courses in digital image processing. Tailoring the contents to the scope of individual semester courses is also an attempt to provide affordable (and "backpack-compatible") textbooks without comprimising the quality and depth of content.

One approach to learning a new language is to become conversant in the core vocabulary and to start using it right away. At first, you may only know how to ask for directions, order coffee, and so on, but once you become confident with the core, you will start engaging others in "conversations" and rapidly learn how to get things done. This step-by-step approach works equally well in many areas of science and engineering.

In this first volume, ostentatiously titled *Fundamental Techniques*, we have attempted to compile the core "vocabulary" of digital image processing, starting from the basic concepts and elementary properties of digital images through simple statistics and point operations, fundamental filtering techniques, localization of edges and contours, and basic operations on color images. Mastering these most commonly used techniques and algorithms will enable you to start being productive right away.

The second volume of this series (*Core Algorithms*) extends the presented material, being devoted to slightly more advanced techniques and algorithms that are, nevertheless, part of the standard image processing toolbox. A forthcoming third volume (*Advanced Techniques*) will extend this series and add important material beyond the elementary level for an advanced undergraduate or even graduate course.

Math, Algorithms, and "Real" Code

While we always concentrate on practical applications and working implementations, we do so without glossing over the important formal details and mathematics necessary for a deeper understanding of the algorithms. In preparing this text, we started from the premise that simply creating a recipe book of imaging solutions would not provide the deeper understanding needed to apply these techniques to novel problems. Instead, our solutions typically develop stepwise along three different perspectives: (a) in mathematical form, (b) as abstract, pseudocode algorithms, and (c) as complete implementations in a real programming language. We use a common and consistent notation throughout to intertwine all three perspectives, thus providing multiple but linked views of the problem and its solution.

Software

The implementations in this series of texts are all based on Java and ImageJ, a widely used programmer-extensible imaging system developed, maintained, and distributed by Wayne Rasband of the National Institutes of Health (NIH).[1] ImageJ is implemented completely in Java and therefore runs on all major platforms. It is widely used because its "plugin"-based architecture enables it to be easily extended. Although all examples run in ImageJ, they have been specifically designed to be easily ported to other environments and programming languages.

We chose Java as an implementation language because it is elegant, portable, familiar to many computing students, and more efficient than commonly thought. Although it may not be the fastest environment for numerical processing of raster images, we think that Java has great advantages when it comes to dynamic data structures and compile-time debugging. Note, however, that we use Java purely as an instructional vehicle because precise semantics are needed and, thus, everything presented here could be easily implemented in almost any other modern programming language. Although we stress the clarity and readability of our software, this is certainly not a book series on Java programming nor does it serve as a reference manual for ImageJ.

---

[1] http://rsb.info.nih.gov/ij/.

Online Resources

The authors maintain a Website for this text that provides supplementary materials, including the complete Java source code for the examples, the test images used in the figures, and corrections. Visit this site at

www.imagingbook.com

Additional materials are available for educators, including a complete set of figures, tables, and mathematical elements shown in the text, in a format suitable for easy inclusion in presentations and course notes. Comments, questions, and corrections are welcome and should be addressed to

imagingbook@gmail.com

Acknowledgements

As with its predecessors, this book would not have been possible without the understanding and steady support of our families. Thanks go to Wayne Rasband at NIH for developing and refining ImageJ and for his truly outstanding support of the growing user community. We appreciate the contribution from many careful readers who have contacted us to suggest new topics, recommend alternative solutions, or suggested corrections. Finally, we are grateful to Wayne Wheeler for initiating this book series and Catherine Brett and her colleagues at Springer's UK and New York offices for their professional support.

Hagenberg, Austria / Washington DC, USA
July 2008

Online Resources

The authors maintain a *Website* for the text that [illegible] including the [illegible] and corrections [illegible]

[illegible]

[illegible]

Contents

1 Digital Images

For a long time, using a computer to manipulate a digital image (i. e., digital image processing) was something performed by only a relatively small group of specialists who had access to expensive equipment. Usually this combination of specialists and equipment was only to be found in research labs, and so the field of digital image processing has its roots in industry and academia. It was not that many years ago that digitizing a photo and saving it to a file on a computer was a time-consuming task. This is perhaps difficult to imagine given today's powerful hardware and operating system level support for all types of digital media, but it is always sobering to remember that "personal" computers in the early 1990s were not powerful enough to even load into main memory a single image from a typical digital camera of today. Now, the combination of a powerful computer on every desktop and the fact that nearly everyone has some type of device for digital image acquisition, be it their cell phone camera, digital camera, or scanner, has resulted in a plethora of digital images and, consequently, for many, digital image processing has become as common as word processing. Powerful hardware and software packages have made it possible for everyone to manipulate digital images and videos.

All of these developments have resulted in a large community that works productively with digital images while having only a basic understanding of the underlying mechanics. And for the typical consumer merely wanting to create a digital archive of vacation photos, a deeper understanding is not required, just as a deep understanding of the combustion engine is unnecessary to successfully drive a car.

Today's IT professionals, however, must be more than simply familiar with

W. Burger, M.J. Burge, *Principles of Digital Image Processing*, Undergraduate Topics in Computer Science, DOI 10.1007/978-1-84800-191-6_1, © Springer-Verlag London Limited, 2009

digital image processing. They are expected to be able to knowledgeably manipulate images and related digital media and, in the same way, software engineers and computer scientists are increasingly confronted with developing programs, databases, and related systems that must correctly deal with digital images. The simple lack of practical experience with this type of material, combined with an often unclear understanding of its basic foundations and a tendency to underestimate its difficulties, frequently leads to inefficient solutions, costly errors, and personal frustration.

1.1 Programming with Images

Even though the term "image processing" is often used interchangeably with that of "image editing", we introduce the following more precise definitions. Digital image editing, or as it is sometimes referred to, digital imaging, is the manipulation of digital images using an existing software application such as Adobe Photoshop or Corel Paint Digital image processing, on the other hand, is the conception, design, development, and enhancement of digital imaging programs.

Modern programming environments, with their extensive APIs (application programming interfaces), make practically every aspect of computing, be it networking, databases, graphics, sound, or imaging, easily available to nonspecialists. The possibility of developing a program that can reach into an image and manipulate the individual elements at its very core is fascinating and seductive. You will discover that with the right knowledge, an image becomes ultimately no more than a simple array of values, and that with the right tools you can manipulate in any way imaginable.

Computer graphics, in contrast to digital image processing, concentrates on the *synthesis* of digital images from geometrical descriptions such as three-dimensional object models [14,16,41]. While graphics professionals today tend to be interested in topics such as realism and, especially in terms of computer games, rendering speed, the field does draw on a number of methods that originate in image processing, such as image transformation (morphing), reconstruction of 3D models from image data, and specialized techniques such as image-based and non-photorealistic rendering [33,42]. Similarly, image processing makes use of a number of ideas that have their origin in computational geometry and computer graphics, such as volumetric (voxel) models in medical image processing. The two fields perhaps work closest when it comes to digital post-production of film and video and the creation of special effects [43]. This book provides a thorough grounding in the effective processing of not only images but also sequences of images; that is, videos.

Digital images are the central theme of this book, and unlike just a few years ago, this term is now so commonly used that there is really no reason to explain it further. Yet, this book is not about all types of digital images, and instead it focuses on *raster* images that are made up of *picture elements*, more commonly known as *pixels*, arranged in a regular rectangular grid.

Every day, people work with a large variety of digital raster images such as color photographs of people and landscapes, grayscale scans of printed documents, building plans, faxed documents, screenshots, medical images such as x-rays and ultrasounds, and a multitude of others (Fig. 1.1). Despite all the different sources for these images, they are all, as a rule, ultimately represented as rectangular ordered arrays of image elements.

1.2 Image Acquisition

The process by which a scene becomes a digital image is varied and complicated, and, in most cases, the images you work with will already be in digital form, so we only outline here the essential stages in the process. As most image acquisition methods are essentially variations on the classical optical camera, we will begin by examining it in more detail.

1.2.1 The Pinhole Camera Model

The pinhole camera is one of the simplest camera models and has been in use since the 13th century, when it was known as the "Camera Obscura". While pinhole cameras have no practical use today except to hobbyists, they are a useful model for understanding the essential optical components of a simple camera.

The pinhole camera consists of a closed box with a small opening on the front side through which light enters, forming an image on the opposing wall. The light forms a smaller, inverted image of the scene (Fig. 1.2).

Perspective transformation

The geometric properties of the pinhole camera are very simple. The optical axis runs through the pinhole perpendicular to the image plane. We assume a visible object (the cactus in Fig. 1.2) located at a horizontal distance Z from the pinhole and vertical distance Y from the optical axis. The height of the projection y is determined by two parameters: the (fixed) depth of the camera box f and the distance Z of the object from the origin of the coordinate system. By matching similar triangles we obtain the relations

$$y = -f\frac{Y}{Z} \qquad \text{and} \qquad x = -f\frac{X}{Z} \tag{1.1}$$

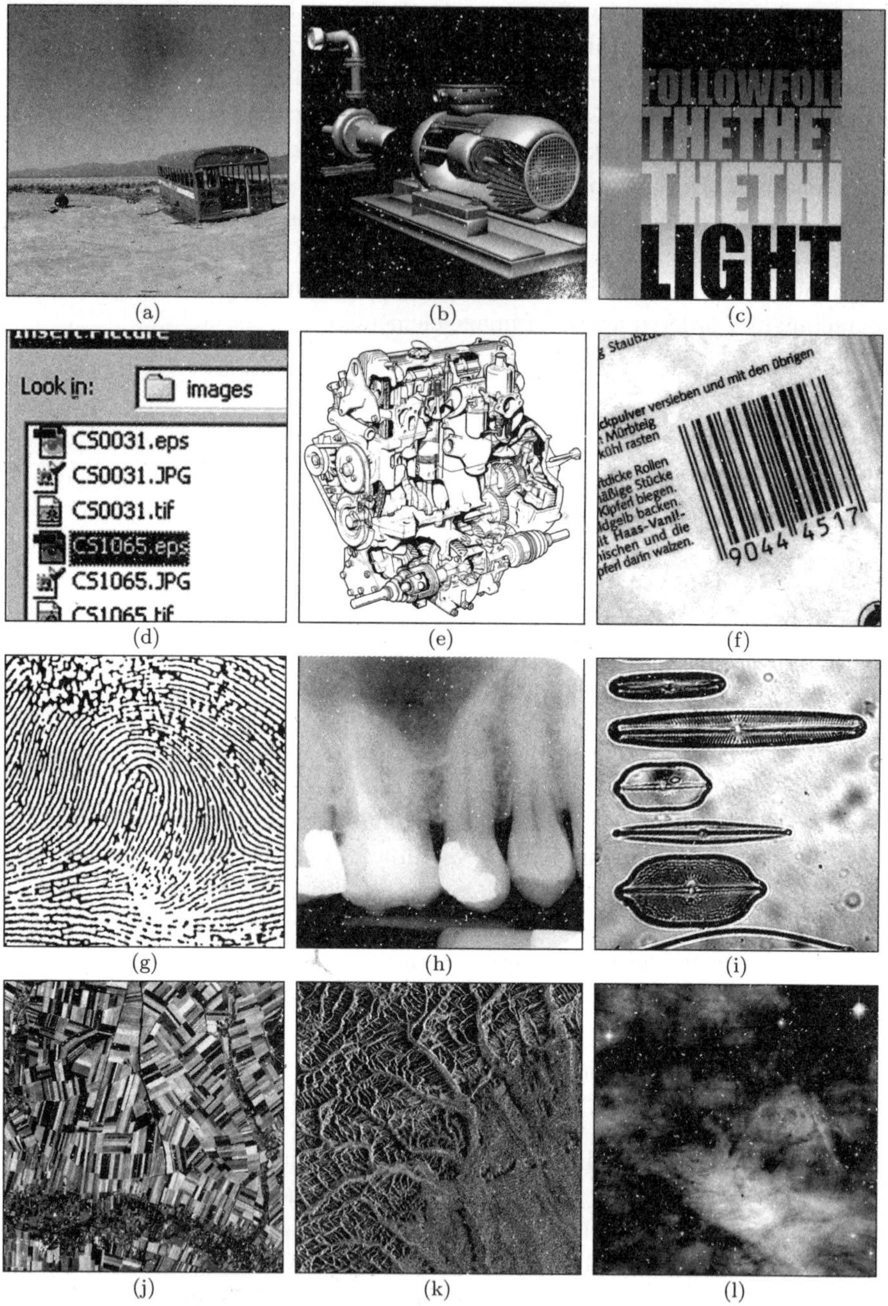

Figure 1.1 Digital images: natural landscape (a), synthetically generated scene (b), poster graphic (c), computer screenshot (d), black and white illustration (e), barcode (f), fingerprint (g), x-ray (h), microscope slide (i), satellite image (j), synthetic radar image (k), astronomical object (l).

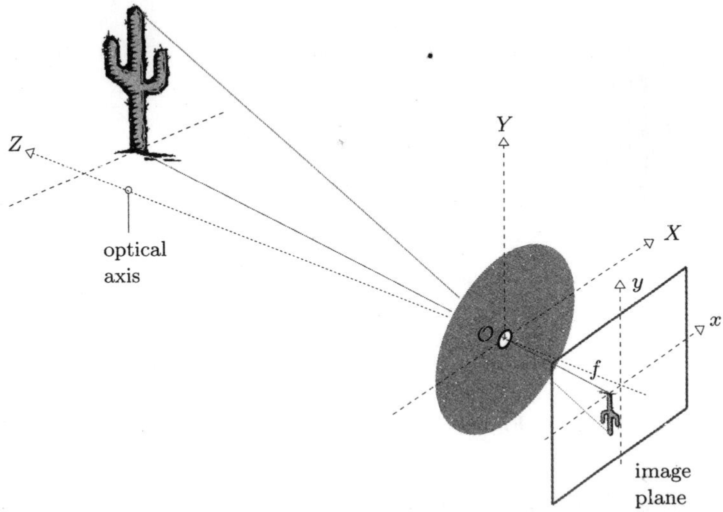

Figure 1.2 Geometry of the pinhole camera. The pinhole opening serves as the origin ($\mathcal{O}$) of the three-dimensional coordinate system (X, Y, Z) for the objects in the scene. The optical axis, which runs through the opening, is the Z axis of this coordinate system. A separate two-dimensional coordinate system (x, y) describes the projection points on the image plane. The distance f ("focal length") between the opening and the image plane determines the scale of the projection.

between the 3D object coordinates X, Y, Z and the corresponding image coordinates x, y for a given focal length f. Obviously, the scale of the resulting image changes in proportion to the distance f in a way similar to how the focal length determines image magnification in an everyday camera. For a fixed scene, a small f (i. e., short focal length) results in a small image and a large viewing angle, just as occurs when a wide-angle lens is used. In contrast, increasing the "focal length" f results in a larger image and a smaller viewing angle, analogous to the effect of a telephoto lens. The negative sign in Eqn. (1.1) means that the projected image is flipped in the horizontal and vertical directions, i. e., it is rotated by 180°. Equation (1.1) describes what is commonly known as the "perspective transformation"[1] from 3D to a 2D image coordinates. Important properties of this theoretical model are, among others, that straight lines in 3D space always map to straight lines in the 2D projections and that circles appear as ellipses.

[1] It is hard to imagine today that the rules of perspective geometry, while known to the ancient mathematicians, were only rediscovered in 1430 by the Renaissance painter Brunoleschi.

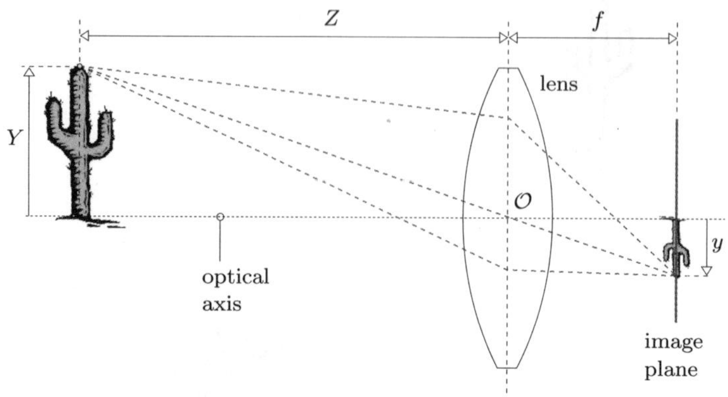

Figure 1.3 The thin lens model.

1.2.2 The "Thin" Lens Model

While the simple geometry of the pinhole camera makes it useful for understanding its basic principles, it is never really used in practice. One of the problems with the pinhole camera is that it requires a very small opening to produce a sharp image. This in turn severely limits the amount of light passed through and thus leads to extremely long exposure times. In reality, glass lenses or systems of optical lenses are used whose optical properties are greatly superior in many aspects, but of course are also much more complex. We can still make our model more realistic, without unduly increasing its complexity, by replacing the pinhole with a "thin lens" as shown in Fig. 1.3.

In this model, the lens is assumed to be symmetric and infinitely thin, such that all light rays passing through it are refracted at a virtual plane in the middle of the lens. The resulting image geometry is practically the same as that of the pinhole camera. This model is not sufficiently complex to encompass the physical details of actual lens systems, such as geometrical distortions and the distinct refraction properties of different colors. So while this simple model suffices for our purposes (that is, understanding the basic mechanics of image acquisition), much more detailed models incorporating these additional complexities can be found in the literature (see, for example, [24]).

1.2.3 Going Digital

What is projected on the image plane of our camera is essentially a two-dimensional, time-dependent, continuous distribution of light energy. In order to obtain a "digital snapshot" of this continuously changing light distribution for processing it on our computer, three main steps are necessary:

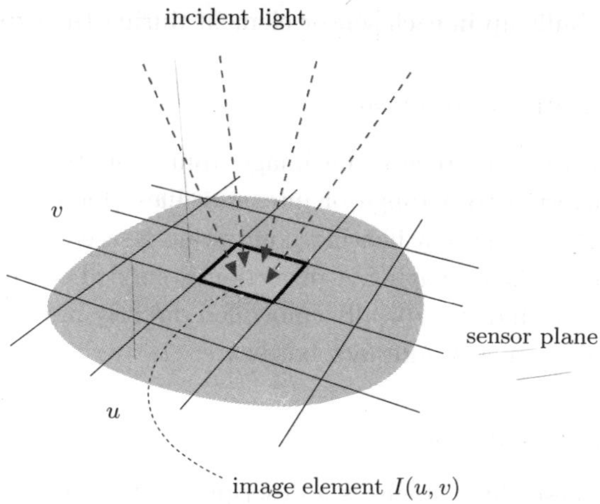

Figure 1.4 The geometry of the sensor elements is directly responsible for the spatial sampling of the continuous image. In the simplest case, a plane of sensor elements are arranged in an evenly spaced raster, and each element measures the amount of light that falls on it.

1. The continuous light distribution must be *spatially sampled.*
2. This resulting "discrete" function must then be *sampled in the time domain* to create a single (still) image.
3. Finally, the resulting values must be *quantized* to a finite set of numeric values so that they are representable within the computer.

Step 1: Spatial sampling

The spatial sampling of an image (that is, the conversion of the continuous signal to its discrete representation) depends on the geometry of the sensor elements of the acquisition device (e. g., a digital or video camera). The individual sensor elements are usually arranged as a rectangular array on the sensor plane (Fig. 1.4). Other types of image sensors, which include hexagonal elements and circular sensor structures, can be found in specialized camera products.

Step 2: Temporal sampling

Temporal sampling is carried out by measuring at regular intervals the amount of light incident on each individual sensor element. The CCD[2] or CMOS[3] sensor in a digital camera does this by triggering an electrical charging process,

[2] Charge-coupled device.

[3] Complementary metal oxyde semiconductor.

induced by the continuous stream of photons, and then measuring the amount of charge that built up in each sensor element during the exposure time.

Step 3: Quantization of pixel values

In order to store and process the image values on the computer they are commonly converted to a range of integer values (for example, $256 = 2^8$ or $4096 = 2^{12}$). Occasionally a floating-point scale is used in professional applications such as medical imaging. Conversion is carried out using an analog to digital converter, which is typically embedded directly in the sensor electronics or is performed by special interface hardware.

Images as discrete functions

The result of these three stages is a description of the image in the form of a two-dimensional, ordered matrix of integers (Fig. 1.5). Stated more formally, a digital image I is a two-dimensional function of integer coordinates $\mathbb{N} \times \mathbb{N}$ that maps to a range of possible image (pixel) values $\mathbb{P}$, such that

$$I(u, v) \in \mathbb{P} \quad \text{and} \quad u, v \in \mathbb{N}.$$

Now we are ready to transfer the image to our computer and save, compress, store or manipulate it in any way we wish. At this point, it is no longer important to us how the image originated since it is now a simple two-dimensional array of numbers. But before moving on, we need a few more important definitions.

1.2.4 Image Size and Resolution

In the following, we assume rectangular images, and while that is a relatively safe assumption, exceptions do exist. The *size* of an image is determined directly from the *width* M (number of columns) and the *height* N (number of rows) of the image matrix I.

The *resolution* of an image specifies the spatial dimensions of the image in the real world and is given as the number of image elements per measurement; for example, *dots per inch* (dpi) or *lines per inch* (lpi) for print production, or in pixels per kilometer for satellite images. In most cases, the resolution of an image is the same in the horizontal and vertical directions, which means that the image elements are square. Note that this is not always the case as, for example, the image sensors of most current video cameras have non-square pixels!

The spatial resolution of an image may not be relevant in many basic image processing steps, such as point operations or filters. Precise resolution

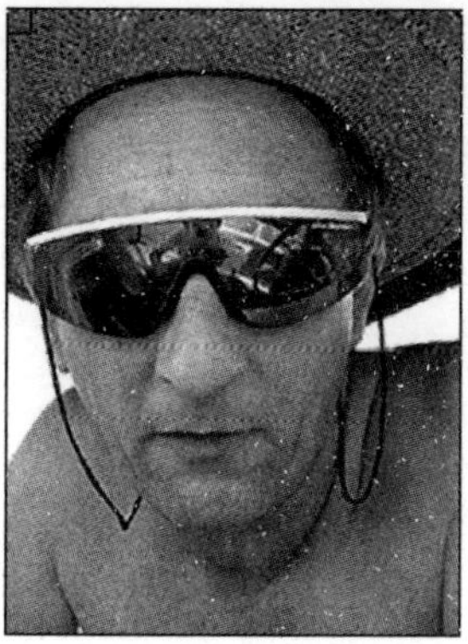

$\longrightarrow$

148	123	52	107	123	162	172	123	64	89	···
147	130	92	95	98	130	171	155	169	163	···
141	118	121	148	117	107	144	137	136	134	···
82	106	93	172	149	131	138	114	113	129	···
57	101	72	54	109	111	104	135	106	125	···
138	135	114	82	121	110	34	76	101	111	···
138	102	128	159	168	147	116	129	124	117	···
113	89	89	109	106	126	114	150	164	145	···
120	121	123	87	85	70	119	64	79	127	···
145	141	143	134	111	124	117	113	64	112	···
⋮	⋮	⋮	⋮	⋮	⋮	⋮	⋮	⋮	⋮	

$F(x, y)$ $\qquad$ $I(u, v)$

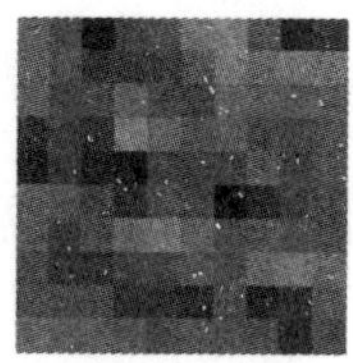

Figure 1.5 Transformation of a continuous intensity function $F(x, y)$ to a discrete digital image $I(u, v)$. The picture below shows the corresponding detail of the discrete intensity image.

information is, however, important in cases where geometrical elements such as circles need to be drawn on an image or when distances within an image need to be measured. For these reasons, most image formats and software systems designed for professional applications rely on precise information about image resolution.

1.2.5 Image Coordinate System

In order to know which position on the image corresponds to which image element, we need to impose a coordinate system. Contrary to normal mathematical conventions, in image processing the coordinate system is usually flipped in the vertical direction; that is, the y-coordinate runs from top to bottom and the origin lies in the upper left corner (Fig. 1.6). While this system has no practical or theoretical advantage, and in fact may be a bit confusing in the context of geometrical transformations, it is used almost without exception in imaging software systems. The system supposedly has its roots in the original design of television broadcast systems, where the picture rows are numbered along the vertical deflection of the electron beam, which moves from the top to the bottom of the screen. We start the numbering of rows and columns at

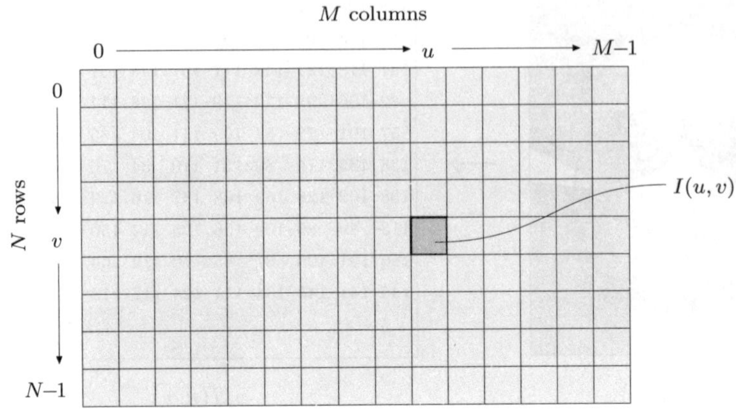

Figure 1.6 Image coordinates. In digital image processing, it is common to use a coordinate system where the origin ($u = 0$, $v = 0$) lies in the upper left corner. The coordinates u, v represent the columns and the rows of the image, respectively. For an image with dimensions $M \times N$, the maximum column number is $u_{\max} = M-1$ and the maximum row number is $v_{\max} = N-1$.

zero for practical reasons, since in Java array indexing also begins at zero.

1.2.6 Pixel Values

The information within an image element depends on the data type used to represent it. Pixel values are practically always binary words of length k so that a pixel can represent any of 2^k different values. The value k is called the bit depth (or just "depth") of the image. The exact bit-level layout of an individual pixel depends on the kind of image; for example, binary, grayscale, or RGB color. The properties of some common image types are summarized below (also see Table 1.1).

Grayscale images (intensity images)

The image data in a grayscale image consist of a single channel that represents the intensity, brightness, or density of the image. In most cases, only positive values make sense, as the numbers represent the intensity of light energy or density of film and thus cannot be negative, so typically whole integers in the range of $[0 \ldots 2^k{-}1]$ are used. For example, a typical grayscale image uses $k = 8$ bits (1 byte) per pixel and intensity values in the range of $[0 \ldots 255]$, where the value 0 represents the minimum brightness (black) and 255 the maximum brightness (white).

For many professional photography and print applications, as well as in medicine and astronomy, 8 bits per pixel is not sufficient. Image depths of 12,

Table 1.1 Bit depths of common image types and typical application domains.

Grayscale (Intensity Images):

Chan.	*Bits/Pix.*	*Range*	*Use*
1	1	0...1	Binary image: document, illustration, fax
1	8	0...255	Universal: photo, scan, print
1	12	0...4095	High quality: photo, scan, print
1	14	0...16383	Professional: photo, scan, print
1	16	0...65535	Highest quality: medicine, astronomy

Color Images:

Chan.	*Bits/Pix.*	*Range*	*Use*
3	24	$[0...255]^3$	RGB, universal: photo, scan, print
3	36	$[0...4095]^3$	RGB, high quality: photo, scan, print
3	42	$[0...16383]^3$	RGB, professional: photo, scan, print
4	32	$[0...255]^4$	CMYK, digital prepress

Special Images:

Chan.	*Bits/Pix.*	*Range*	*Use*
1	16	$-32768...32767$	Integer values pos./neg., increased range
1	32	$\pm 3.4 \cdot 10^{38}$	Floating-point values: medicine, astronomy
1	64	$\pm 1.8 \cdot 10^{308}$	Floating-point values: internal processing

14, and even 16 bits are often encountered in these domains. Note that bit depth usually refers to the number of bits used to represent one color component, not the number of bits needed to represent an entire color pixel. For example, an RGB-encoded color image with an 8-bit depth would require 8 bits for each channel for a total of 24 bits, while the same image with a 12-bit depth would require a total of 36 bits.

Binary images

Binary images are a special type of intensity image where pixels can only take on one of two values, black or white. These values are typically encoded using a single bit (0/1) per pixel. Binary images are often used for representing line graphics, archiving documents, encoding fax transmissions, and of course in electronic printing.

Color images

Most color images are based on the primary colors red, green, and blue (RGB), typically making use of 8 bits for each color component. In these color images, each pixel requires $3 \times 8 = 24$ bits to encode all three components, and the range of each individual color component is $[0 \ldots 255]$. As with intensity images, color images with 30, 36, and 42 bits per pixel are commonly used in professional applications. Finally, while most color images contain three components, images with four or more color components are common in most prepress applications, typically based on the subtractive CMYK (**C**yan-**M**agenta-**Y**ellow-Blac**k**) color model (see Ch. 8).

Indexed or *palette* images constitute a very special class of color image. The difference between an indexed image and a *true color* image is the number of different colors (fewer for an indexed image) that can be used in a particular image. In an indexed image, the pixel values are only indices (with a maximum of 8 bits) onto a specific table of selected full-color values (see Sec. 8.1.1).

Special images

Special images are required if none of the above standard formats is sufficient for representing the image values. Two common examples of special images are those with negative values and those with floating-point values. Images with negative values arise during image-processing steps, such as filtering for edge detection (see Sec. 6.2.2), and images with floating-point values are often found in medical, biological or astronomical applications, where extended numerical range and precision are required. These special formats are mostly application-specific and thus may be difficult to use with standard image-processing tools.

1.3 Image File Formats

While in this book we almost always consider image data as being already in the form of a two-dimensional array—ready to be accessed by a program—, in practice image data must first be loaded into memory from a file. Files provide the essential mechanism for storing, archiving, and exchanging image data, and the choice of the correct file format is an important decision. In the early days of digital image processing (that is, before around 1985), most software developers created a new custom file format for almost every new application they developed. The result was a chaotic jumble of incompatible file formats that for a long time limited the practical sharing of images between research groups. Today there exist a wide range of standardized file formats, and developers can almost always find at least one existing format that is suitable for their application. Using standardized file formats vastly increases the ease with which images can be exchanged and the likelihood that the images will be readable by other software in the longterm. Yet for many projects the selection of the right file format is not always simple, and compromises must be made. The following are a few of the typical criteria that need to be considered when selecting an appropriate file format:

Type of image: These include black and white images, grayscale images, scans from documents, color images, color graphics, and special images such as those using floating-point image data. In many applications, such as satellite imagery, the maximum image size is also an important factor.

Storage size and compression: Are the storage requirements of the file a potential problem, and is the image compression method, especially when considering *lossy compression*, appropriate?

Compatibility: How important is the exchange of image data? And for archives, how important is the long-term machine readability of the data?

Application domain: In which domain will the image data be mainly used? Are they intended for print, Web, film, computer graphics, medicine, or astronomy?

1.3.1 Raster versus Vector Data

In the following, we will deal exclusively with file formats for storing *raster images*; that is, images that contain pixel values arranged in a regular matrix using discrete coordinates. In contrast, *vector graphics* represent geometric objects using continuous coordinates, which are only rasterized once they need to be displayed on a physical device such as a monitor or printer.

A number of standardized file formats exist for vector images, such as the ANSI/ISO standardformat CGM (Computer Graphics Metafile), SVG (Scalable Vector Graphics)[4] as well as proprietary formats such as DXF (Drawing Exchange Format from AutoDesk), AI (Adobe Illustrator), PICT (QuickDraw Graphics Metafilefrom Apple) and WMF/EMF (Windows Metafile and Enhanced Metafile from Microsoft). Most of these formats can contain both vector data and raster images in the same file. The PS (PostScript) and EPS (Encapsulated PostScript) formats from Adobe as well as the PDF (Portable Document Format) also offer this possibility, though they are usually used for printer output and archival purposes.[5]

1.3.2 Tagged Image File Format (TIFF)

This is a widely used and flexible file format designed to meet the professional needs of diverse fields. It was originally developed by Aldus and later extended

[4] www.w3.org/TR/SVG/.

[5] Special variations of PS, EPS, and PDF files are also used as (editable) exchange formats for raster and vector data; for example, both Adobe's Photoshop (Photoshop-EPS) and Illustrator (AI).

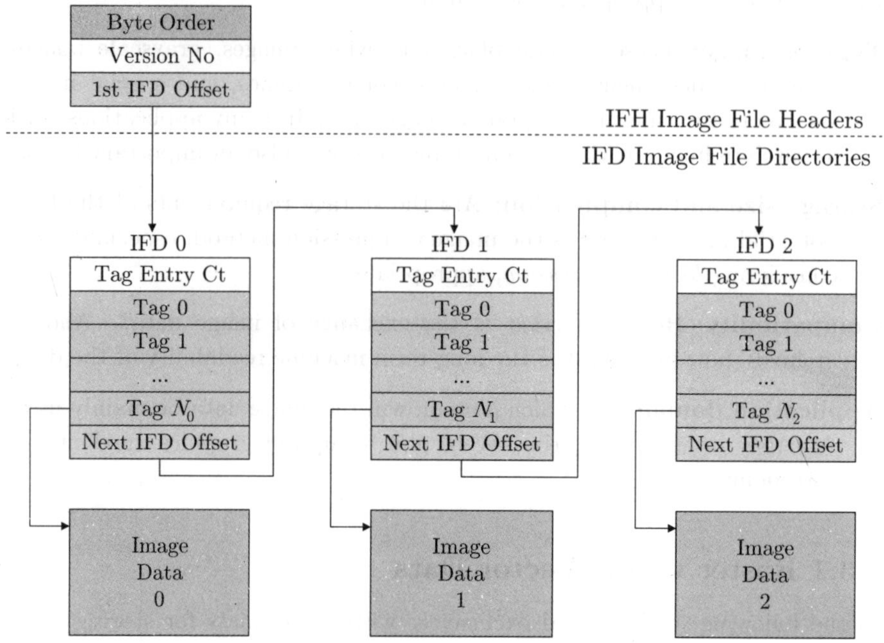

Figure 1.7 Structure of a typical TIFF file. A TIFF file consists of a header and a linked list of image objects, three in this example. Each image object consists of a list of "tags" with their corresponding entries followed by a pointer to the actual image data.

by Microsoft and currently Adobe. The format supports a range of grayscale, indexed, and true color images, but also special image types with large-depth integer and floating-point elements. A TIFF file can contain a number of images with different properties. The TIFF specification provides a range of different compression methods (LZW, ZIP, CCITT, and JPEG) and color spaces, so that it is possible, for example, to store a number of variations of an image in different sizes and representations together in a single TIFF file. The flexibility of TIFF has made it an almost universal exchange format that is widely used in archiving documents, scientific applications, digital photography, and digital video production.

The strength of this image format lies within its architecture (Fig. 1.7), which enables new image types and information blocks to be created by defining new "tags". In this flexibility also lies the weakness of the format, namely that proprietary tags are not always supported and so the "unsupported tag" error is sometimes still encountered when loading TIFF files. ImageJ also reads only a few uncompressed variations of TIFF formats,[6] and bear in mind that most

[6] The `ImageIO` plugin offers support for a wider range of TIFF formats.

popular Web browsers currently do not support TIFF either.

1.3.3 Graphics Interchange Format (GIF)

The Graphics Interchange Format (GIF) was originally designed by CompuServe in 1986 to efficiently encode the rich line graphics used in their dial-up Bulletin Board System (BBS). It has since grown into one of the most widely used formats for representing images on the Web. This popularity is largely due to its early support for indexed color at multiple bit depths, LZW compression, interlaced image loading, and ability to encode simple animations by storing a number of images in a single file for later sequential display.

GIF is essentially an indexed image file format designed for color and gray scale images with a maximum depth of 8 bits and consequently it does not support true color images. It offers efficient support for encoding palettes containing from 2 to 256 colors, one of which can be marked for transparency. GIF supports color palletes in the range of $2 \ldots 256$, enabling pixels to be encoded using fewer bits. As an example, the pixels of an image using 16 unique colors require only 4 bits to store the 16 possible color values $[0 \ldots 15]$. This means that instead of storing each pixel using one byte, as done in other bitmap formats, GIF can encode two 4-bit pixels into each 8-bit byte. This results in a 50% storage reduction over the standard 8-bit indexed color bitmap format.

The GIF file format is designed to efficiently encode "flat" or "iconic" images consisting of large areas of the same color. It uses a lossless color quantization (see Vol. 2 [6, Sec. 5]) as well as lossless LZW compression to efficiently encode large areas of the same color. Despite the popularity of the format, when developing new software, the PNG format, presented in the next section, should be preferred, as it outperforms GIF by almost every metric.

1.3.4 Portable Network Graphics (PNG)

PNG (pronounced "ping") was originally developed as a replacement for the GIF file format when licensing issues[7] arose because of its use of LZW compression. It was designed as a universal image format especially for use on the Internet, and, as such, PNG supports three different types of images:

- true color (with up to 3×16 bits/pixel)
- grayscale (with up to 16 bits/pixel)
- indexed (with up to 256 colors)

[7] Unisys's U.S. LZW Patent No. 4,558,302 expired on June 20, 2003.

Additionally, PNG includes an *alpha* channel for transparency with a maximum depth of 16 bits. In comparison, the transparency channel of a GIF image is only a single bit deep. While the format only supports a single image per file, it is exceptional in that it allows images of up to $2^{30} \times 2^{30}$ pixels. The format supports lossless compression by means of a variation of PKZIP (Phil Katz's ZIP). No lossy compression is available, as PNG was not designed as a replacement for JPEG. Ultimately the PNG format meets or exceeds the capabilities of the GIF format in every way except GIF's ability to include multiple images in a single file to create simple animations. Currently, PNG should be considered the format of choice for representing uncompressed, lossless, true color images for use on the Web.

1.3.5 JPEG

The JPEG standard defines a compression method for continuous grayscale and color images, such as those that would arise from nature photography. The format was developed by the Joint Photographic Experts Group (JPEG)[8] with the goal of achieving an average data reduction of a factor of 1:16 and was established in 1990 as ISO Standard IS-10918. Today it is the most widely used image file format. In practice, JPEG achieves, depending on the application, compression in the order of 1 bit per pixel (that is, a compression factor of around 1:25) when compressing 24-bit color images to an acceptable quality for viewing. The JPEG standard supports images with up to 256 color components, and what has become increasingly important is its support for CMYK images (see Sec. 8.2.5).

In the case of RGB images, the core of the algorithm consists of three main steps:

1. **Color conversion and down sampling:** A color transformation from RGB into the YC_bC_r space (see Sec. 8.2.4) is used to separate the actual color components from the brightness Y component. Since the human visual system is less sensitive to rapid changes in color, it is possible to compress the color components more, resulting in a significant data reduction, without a subjective loss in image quality.

2. **Cosine transform and quantization in frequency space:** The image is divided up into a regular grid of 8 blocks, and for each independent block, the frequency spectrum is computed using the discrete cosine transformation (see Vol. 2 [6, Ch. 9]). Next, the 64 spectral coefficients of each block are quantized into a quantization table. The size of this table largely determines the eventual compression ratio, and therefore the visual quality,

[8] www.jpeg.org.

of the image. In general, the high frequency coefficients, which are essential for the "sharpness" of the image, are reduced most during this step. During decompression these high frequency values will be approximated by computed values.

3. **Lossless compression:** Finally, the quantized spectral components data stream is again compressed using a lossless method, such as arithmetic or Huffman encoding, in order to remove the last remaining redundancy in the data stream.

In addition to the "baseline" algorithm, several other variants are provided, including a (rarely used) uncompressed version. The JPEG compression method combines a number of different compression methods and is quite complex in its entirety [30]. Implementing even the baseline version is nontrivial, so application support for JPEG increased sharply once the Independent JPEG Group (IJG)[9] made available a reference implementation of the JPEG algorithm in 1991.

Drawbacks of the JPEG compression algorithm include its limitation to 8-bit images, its poor performance on non-photographic images such as line art (for which it was not designed), its handling of abrupt transitions within an image, and the striking artifacts caused by the 8×8 pixel blocks at high compression rates. Figure 1.9 shows the results of compressing a section of a grayscale image using different quality factors (Photoshop $Q_{\mathrm{JPG}} = 10, 5, 1$).

JFIF file format

Despite common usage, JPEG is *not* a file format; it is "only" a method of compressing image data. The actual JPEG standard only specifies the JPEG codec (compressor and decompressor) and by design leaves the wrapping, or file format, undefined.[10] (Fig. 1.8). What is normally referred to as a JPEG *file* is almost always an instance of a "JPEG File Interchange Format" (JFIF) file, originally developed by Eric Hamilton and the IJG. The JFIF specifies a file format based on the JPEG standard by defining the remaining necessary elements of a file format. The JPEG standard leaves some parts of the codec undefined for generality, and in these cases JFIF makes a specific choice. As an example, in step 1 of the JPEG codec, the specific color space used in the color transformation is not part of the JPEG standard, so it is specified by the JFIF standard. As such, the use of different compression ratios for color and luminance is a practical implementation decision specified by JFIF and is not a part of the actual JPEG codec.

---

[9] www.ijg.org.

[10] To be exact, the JPEG standard only defines how to compress the individual components and the structure of the JPEG stream.

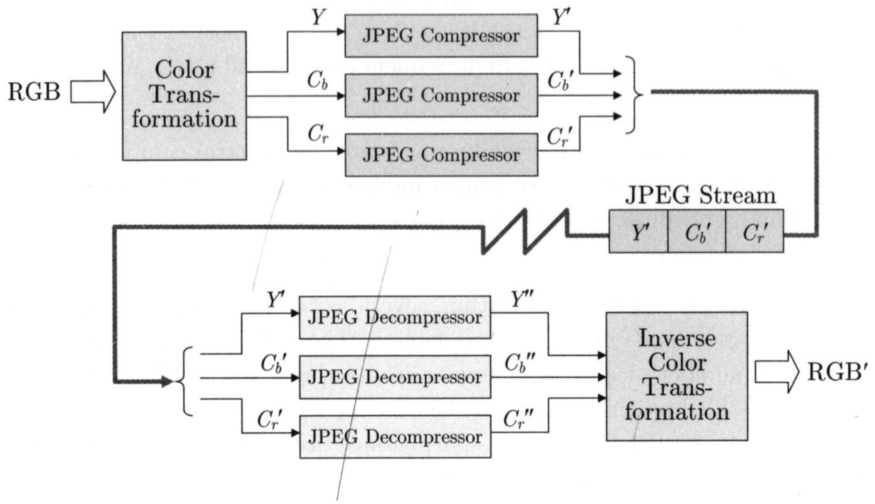

Figure 1.8 JPEG compression of an RGB image. Using a color space transformation, the color components C_b, C_r are separated from the Y luminance component and subjected to a higher rate of compression. Each of the three components are then run independently through the JPEG compression pipeline and are merged into a single JPEG data stream. Decompression follows the same stages in reverse order.

Exchangeable Image File Format (EXIF)

The Exchangeable Image File Format (EXIF) is a variant of the JPEG (JFIF) format designed for storing image data originating on digital cameras, and to that end it supports storing metadata such as the type of camera, date and time, photographic parameters such as aperture and exposure time, as well as geographical (GPS) data. EXIF was developed by the Japan Electronics and Information Technology Industries Association (JEITA) as a part of the DCF[11] guidelines and is used today by practically all manufacturers as the standard format for storing digital images on memory cards. Internally, EXIF uses TIFF to store the metadata information and JPEG to encode a thumbnail preview image. The file structure is designed so that it can be processed by existing JPEG/JFIF readers without a problem.

JPEG-2000

JPEG-2000, which is specified by an ISO-ITU standard ("Coding of Still Pictures"),[12] was designed to overcome some of the better-known weaknesses of the traditional JPEG codec. Among the improvements made in JPEG-2000

[11] Design Rule for Camera File System.
[12] www.jpeg.org/JPEG2000.htm.

(a) Original
(75.08 kB)

(b) $Q_{\text{JPG}} = 10$
(11.40 kB)

(c) $Q_{\text{JPG}} = 5$
(7.24 kB)

(d) $Q_{\text{JPG}} = 1$
(5.52 kB)

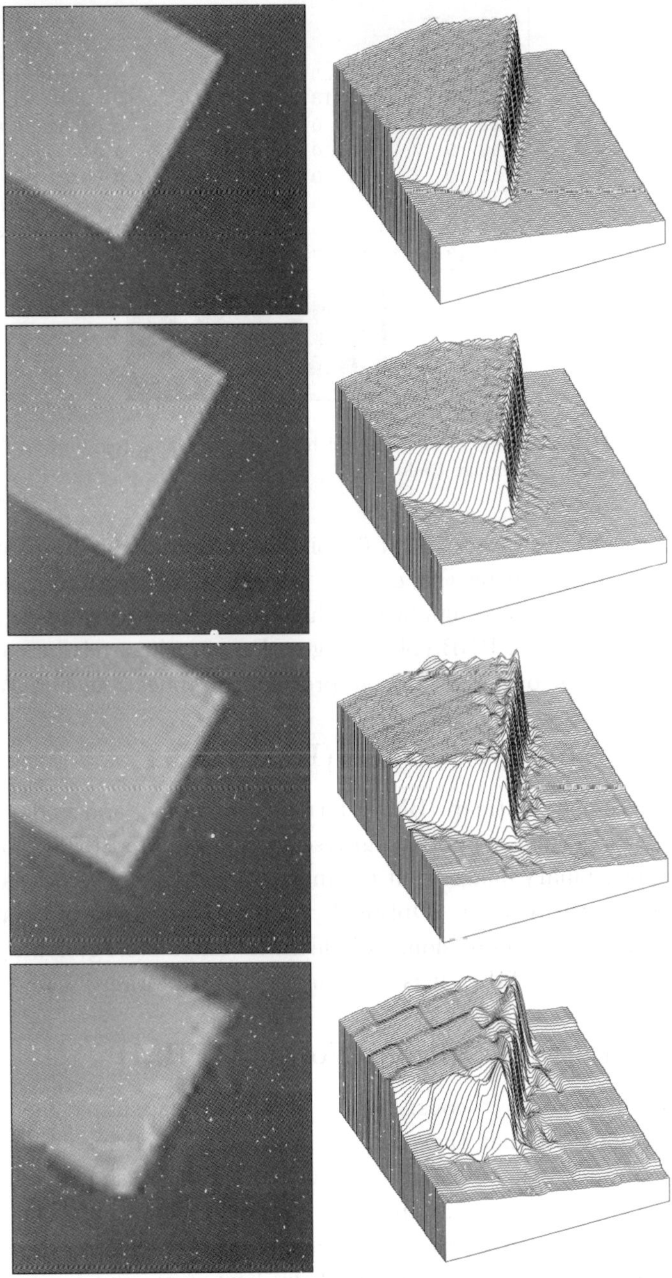

Figure 1.9 Artifacts arising from JPEG compression. A section of the original image (a) and the results of JPEG compression at different quality factors: $Q_{\text{JPG}} = 10$ (b), $Q_{\text{JPG}} = 5$ (c), and $Q_{\text{JPG}} = 1$ (d). In parentheses are the resulting file sizes for the complete (dimensions 274×274) image.

```
P2
# oie.pgm
17 7
255
0 13 13 13 13 13 13 13  0  0  0  0  0  0  0  0  0
0 13  0  0  0  0  0 13  0  7  7  0  0 81 81 81 81
0 13  0  7  7  7  0 13  0  7  7  0  0 81  0  0  0
0 13  0  7  0  7  0 13  0  7  7  0  0 81 81 81  0
0 13  0  7  7  7  0 13  0  7  7  0  0 81  0  0  0
0 13  0  0  0  0  0 13  0  7  7  0  0 81 81 81 81
0 13 13 13 13 13 13 13  0  0  0  0  0  0  0  0  0
```

Figure 1.10 Example of a PGM file in human-readable text format (top) and the corresponding grayscale image (below).

are the use of larger, 64 × 64 pixel blocks and replacement of the discrete cosine transform by the *wavelet* transform. These and other improvements enable it to achieve significantly higher compression ratios than JPEG—up to 0.25 bit/pixel on RGB color images. Despite these advantages, JPEG-2000 is supported by only a few image-processing applications and Web browsers.[13]

1.3.6 Windows Bitmap (BMP)

The Windows Bitmap (BMP) format is a simple, and under Windows widely used, file format supporting grayscale, indexed, and true color images. It also supports binary images, but not in an efficient manner since each pixel is stored using an entire byte. Optionally, the format supports simple lossless, run-length-based compression. While BMP offers storage for a similar range of image types as TIFF, it is a much less flexible format.

1.3.7 Portable Bitmap Format (PBM)

The PBM family[14] consists of a series of very simple file formats that are exceptional in that they can be optionally saved in a human-readable text format that can be easily read in a program or simply edited using a text editor. A simple PGM image is shown in Fig. 1.10. The characters `P2` in the first line indicate that the image is a PGM ("plain") file stored in human-readable format. The next line shows how comments can be inserted directly into the file by beginning the line with the # symbol. Line 3 gives the image's

[13] At this time, ImageJ does not offer JPEG-2000 support.

[14] http://netpbm.sourceforge.net.

dimensions, in this case width 17 and height 7, and line 4 defines the maximum pixel value, in this case 255. The remaining lines give the actual pixel values. This format makes it easy to create and store image data without any explicit imaging API, since it requires only basic text I/O that is available in any programming environment.

In addition, the format supports a much more machine-optimized "raw" output mode in which pixel values are stored as bytes. PBM is widely used under Unix and supports the following formats: PBM (*portable bitmap*) for binary *bitmaps*, PGM (*portable graymap*) for grayscale images, and PNM (*portable any map*) for color images. PGM images can be opened using ImageJ.

1.3.8 Additional File Formats

For most practical applications, one of the following file formats is sufficient: TIFF as a universal format supporting a wide variety of uncompressed images and JPEG/JFIF for digital color photos when storage size is a concern, and there is either PNG or GIF for when an image is destined for use on the Web. In addition, there exist countless other file formats, such as those encountered in legacy applications or in special application areas where they are traditionally used. A few of the more commonly encountered types are:

- **RGB**, a simple format from Silicon Graphics.
- **RAS** (Sun Raster Format), a simple format from Sun Microsystems.
- **TGA** (Truevision Targa File Format) was the first 24-bit file format for PCs. It supports numerous image types with 8- to 32-bit depths and is still used in medicine and biology.
- **XBM/XPM** (X-Windows Bitmap/Pixmap) is a family of ASCII-encoded formats used in X-Windows and is similar to PBM/PGM.

1.3.9 Bits and Bytes

Today, opening, reading, and writing image files is mostly carried out by means of existing software libraries. Yet sometimes you still need to deal with the structure and contents of an image file at the byte level, for instance when you need to read an unsupported file format or when you receive a file where the format of the data is unknown.

Big endian and little endian

In the standard model of a computer, a file consists of a simple sequence of 8-bit bytes, and a byte is the smallest entry that can be read or written to a file. In contrast, the image elements as they are stored in memory are usually larger then a byte; for example, a 32-bit `int` value (= 4 bytes) is used for an RGB color pixel. The problem is that storing the four individual bytes that make up the image data can be done in different ways. In order to correctly recreate the original color pixel, we must naturally know the *order* in which bytes in the file are arranged.

Consider a 32-bit `int` number z with the binary and hexadecimal value[15]

$$z = \underbrace{00010010}_{\substack{12_H \\ \text{(MSB)}}}\,00110100\,01010110\,\underbrace{01111000}_{\substack{78_H \\ \text{(LSB)}}}{}_B = 12345678_H. \qquad (1.2)$$

Then $00010010_B = 12_H$ is the value of the *most significant byte* (MSB) and $01111000_B = 78_H$ the *least significant byte* (LSB). When the individual bytes in the file are arranged in order from MSB to LSB when they are saved, we call the ordering "big endian", and when in the opposite direction, "little endian". Thus the 32-bit value z from Eqn. (1.2) could be stored in one of the following two modes:

Ordering	Byte Sequence	1	2	3	4
Big Endian	MSB → LSB	12_H	34_H	56_H	78_H
Little Endian	LSB → MSB	78_H	56_H	34_H	12_H

Even though correctly ordering the bytes should essentially be the responsibility of the operating and file system, in practice it actually depends on the architecture of the processor.[16] Processors from the Intel family (e.g., x86, Pentium) are traditionally little endian, and processors from other manufacturers (e.g., IBM, MIPS, Motorola, Sun) are big endian.[17] Big endian is also called *network byte ordering* since in the IP protocol the data bytes are arranged in MSB to LSB order during transmission.

To correctly interpret image data with multi-byte pixel values, it is necessary to know the byte ordering used when creating it. In most cases, this is fixed and defined by the file format, but in some file formats, for example TIFF, it is variable and depends on a parameter given in the file header (see Table 1.2).

File headers and signatures

Practically all image file formats contain a data header consisting of important information about the layout of the image data that follows. Values such as the size of the image and the encoding of the pixels are usually present in the

---

[15] The decimal value of z is 305419896.

[16] At least the ordering of the *bits* within a byte is almost universally uniform.

[17] In Java, this problem does not arise since internally all implementations of the *Java Virtual Machine* use big endian ordering.

Table 1.2 Signatures of various image file formats. Most image file formats can be identified by inspecting the first bytes of the file. These byte sequences, or signatures, are listed in hexadecimal (`0x..`) form and as ASCII text (□ indicates a nonprintable character).

Format	*Signature*		*Format*	*Signature*	
PNG	`0x89504e47`	`□PNG`	BMP	`0x424d`	`BM`
JPEG/JFIF	`0xffd8ffe0`	`□□□□`	GIF	`0x4749463839`	`GIF89`
$\text{TIFF}_{\text{little}}$	`0x49492a00`	`II*□`	Photoshop	`0x38425053`	`8BPS`
TIFF_{big}	`0x4d4d002a`	`MM□*`	PS/EPS	`0x25215053`	`%!PS`

file header to make it easier for programmers to allocate the correct amount of memory for the image. The size and structure of this header are usually fixed, but in some formats such as TIFF, the header can contain pointers to additional subheaders.

In order to interpret the information in the header, it is necessary to know the file type. In many cases, this can be determined by the *file name extension* (e.g., `.jpg` or `.tif`), but since these extensions are not standardized and can be changed at any time by the user, they are not a reliable way of determining the file type. Instead, many file types can be identified by their embedded "signature", which is often the first two bytes of the file. Signatures from a number of popular image formats are given in Table 1.2. Most image formats can be determined by inspecting the first few bytes of the file. These bytes, or signatures, are listed in hexadecimal (`0x..`) form and as ASCII text. A PNG file always begins with the 4-byte sequence `0x89`, `0x50`, `0x4e`, `0x47`, which is the "magic number" `0x89` followed by the ASCII sequence "PNG". Sometimes the signature not only identifies the type of image file but also contains information about its encoding; for instance, in TIFF the first two characters are either `II` for "Intel" or `MM` for "Motorola" and indicate the byte ordering (little endian or big endian, respectively) of the image data in the file.

1.4 Exercises

Exercise 1.1

Determine the actual physical measurement in millimeters of an image with 1400 rectangular pixels and a resolution of 72 dpi.

Exercise 1.2

A camera with a focal length of $f = 50$ mm is used to take a photo of a vertical column that is 12 m high and is 95 m away from the camera. Determine its height in the image in mm (a) and the number of pixels (b) assuming the camera has a resolution of 4000 dots per inch (dpi).

Exercise 1.3

The image sensor of a certain digital camera contains 2016×3024 pixels. The geometry of this sensor is identical to that of a traditional 35 mm camera (with an image size of 24×36 mm) except that it is 1.6 times smaller. Compute the resolution of this digital sensor in *dots per inch*.

Exercise 1.4

Assume the camera geometry described in Exercise 1.3 combined with a lens with focal length $f = 50$ mm. What amount of blurring (in pixels) would be caused by a uniform, $0.1°$ horizontal turn of the camera during exposure? Recompute this for $f = 300$ mm. Decide if the extent of the blurring also depends on the distance of the object.

Exercise 1.5

Determine the number of bytes necessary to store an uncompressed binary image of size 4000×3000 pixels.

Exercise 1.6

Determine the number of bytes necessary to store an uncompressed RGB color image of size 640×480 pixels using 8, 10, 12, and 14 bits per color channel.

Exercise 1.7

Given a black and white television with a resolution of 625×512 8-bit pixels and a frame rate of 25 images per second: (a) How may different images can this device ultimately display, and how long would you have to watch it (assuming no sleeping) in order to see every possible image at least once? (b) Perform the same calculation for a color television with 3×8 bits per pixel.

Exercise 1.8

Show that the projection of a 3D straight line in a pinhole camera (assuming perspective projection as defined in Eqn. (1.1)) is again a straight line in the resulting 2D image.

Exercise 1.9

Using Fig. 1.10 as a model, use a text editor to create a PGM file, `disk.pgm`, containing an image of a bright circle. Open your image with ImageJ and then try to find other programs that can open and display the image.

2
ImageJ

Until a few years ago, the image-processing community was a relatively small group of people who either had access to expensive commercial image-processing tools or, out of necessity, developed their own software packages. Usually such home-brew environments started out with small software components for loading and storing images from and to disk files. This was not always easy because often one had to deal with poorly documented or even proprietary file formats. An obvious (and frequent) solution was to simply design a *new* image file format from scratch, usually optimized for a particular field, application, or even a single project, which naturally led to a myriad of different file formats, many of which did not survive and are forgotten today [30, 32]. Nevertheless, writing software for *converting* between all these file formats in the 1980s and early 1990s was an important business that occupied many people. Displaying images on computer screens was similarly difficult, because there was only marginal support by operating systems, APIs, and display hardware, and capturing images or videos into a computer was close to impossible on common hardware. It thus may have taken many weeks or even months before one could do just elementary things with images on a computer and finally do some serious image processing.

Fortunately, the situation is much different today. Only a few common image file formats have survived (see also Sec. 1.3), which are readily handled by many existing tools and software libraries. Most standard APIs for C/C++, Java, and other popular programming languages already come with at least some basic support for working with images and other types of media data. While there is still much development work going on at this level, it makes our

W. Burger, M.J. Burge, *Principles of Digital Image Processing*, Undergraduate Topics in Computer Science, DOI 10.1007/978-1-84800-191-6_2, © Springer-Verlag London Limited, 2009

job a lot easier and, in particular, allows us to focus on the more interesting aspects of digital imaging.

2.1 Image Manipulation and Processing

Traditionally, software for digital imaging has been targeted at either *manipulating* or *processing* images, either for practitioners and designers or software programmers, with quite different requirements.

Software packages for *manipulating* images, such as Adobe Photoshop, Corel Paint and others, usually offer a convenient user interface and a large number of readily available functions and tools for working with images interactively. Sometimes it is possible to extend the standard functionality by writing scripts or adding self-programmed components. For example, Adobe provides a special API[1] for programming Photoshop "plugins" in C++, though this is a nontrivial task and certainly too complex for nonprogrammers.

In contrast to the category of tools above, digital image *processing* software primarily aims at the requirements of algorithm and software developers, scientists, and engineers working with images, where interactivity and ease of use are not the main concerns. Instead, these environments mostly offer comprehensive and well-documented software libraries that facilitate the implementation of new image-processing algorithms, prototypes and working applications. Popular examples are Khoros/VisiQuest,[2] IDL,[3] MatLab,[4] and ImageMagick,[5] among many others. In addition to the support for conventional programming (typically with C/C++), many of these systems provide dedicated scripting languages or visual programming aides that can be used to construct even highly complex processes in a convenient and safe fashion.

In practice, image manipulation and image processing are of course closely related. Although Photoshop, for example, is aimed at image manipulation by nonprogrammers, the software itself implements many traditional image-processing algorithms. The same is true for many Web applications using server-side image processing, such as those based on ImageMagick. Thus image processing is really at the base of any image manipulation software and certainly not an entirely different category.

[1] www.adobe.com/products/photoshop/.
[2] www.accusoft.com/imaging/visiquest/.
[3] www.rsinc.com/idl/.
[4] www.mathworks.com.
[5] www.imagemagick.org.

2.2 ImageJ Overview

ImageJ, the software that is used for this book, is a combination of both worlds discussed above. It offers a set of ready-made tools for viewing and interactive manipulation of images but can also be extended easily by writing new software components in a "real" programming language. ImageJ is implemented entirely in Java and is thus largely platform-independent, running without modification under Windows, MacOS, or Linux. Java's dynamic execution model allows new modules ("plugins") to be written as independent pieces of Java code that can be compiled, loaded, and executed "on the fly" in the running system without the need to even restart ImageJ. This quick turnaround makes ImageJ an ideal platform for developing and testing new image-processing techniques and algorithms. Since Java has become extremely popular as a first programming language in many engineering curricula, it is usually quite easy for students to get started in ImageJ without spending much time to learn another programming language. Also, ImageJ is freely available, so students, instructors, and practitioners can install and use the software legally and without license charges on any computer. ImageJ is thus an ideal platform for education and self-training in digital image processing but is also in regular use for serious research and application development at many laboratories around the world, particularly in biological and medical imaging.

ImageJ was (and still *is*) developed by Wayne Rasband [34] at the U.S. National Institutes of Health (NIH), originally as a substitute for its predecessor, NIH-Image, which was only available for the Apple Macintosh platform. The current version of ImageJ, updates, documentation, the complete source code, test images, and a continuously growing collection of third-party plugins can be downloaded from the ImageJ Website.[6] Installation is simple, with detailed instructions available online, in Werner Bailer's programming tutorial [3], and in the authors' *ImageJ Short Reference* [5].

To give a structured orientation on ImageJ, this short reference[7] is grouped into different task areas and concentrates on the key functionalities. Some specific rarely used functions were deliberately omitted, but they can of course be found in the ImageJ documentation and the (online) source code.

2.2.1 Key Features

As a pure Java application, ImageJ should run on any computer for which a current Java runtime environment (JRE) exists. ImageJ comes with its own Java runtime, so Java need not be installed separately on the computer. Under

[6] http://rsb.info.nih.gov/ij/.

[7] Available at www.imagingbook.com.

Figure 2.1 Wayne Rasband (right), author of ImageJ, at the 1st ImageJ Conference 2006 (picture courtesy of Marc Seil, CRP Henri Tudor, Luxembourg).

the usual restrictions, ImageJ can be run as a Java "applet" within a Web browser, though it is mostly used as a stand-alone application. It is sometimes also used on the server side in the context of Java-based Web applications (see [3] for details). In summary, the key features of ImageJ are:

- A set of ready-to-use, interactive tools for creating, visualizing, editing, processing, analyzing, loading, and storing images, with support for several common file formats. ImageJ also provides "deep" 16-bit integer images, 32-bit floating-point images, and image sequences ("stacks").
- A simple plugin mechanism for extending the core functionality of ImageJ by writing (usually small) pieces of Java code. All coding examples shown in this book are based on such plugins.
- A macro language and the corresponding interpreter, which make it easy to implement larger processing blocks by combining existing functions without any knowledge of Java. Macros are not discussed in this book, but details can be found in ImageJ's online documentation.[8]

2.2.2 Interactive Tools

When ImageJ starts up, it first opens its main window (Fig. 2.2), which includes the following menu entries:

- File: opening, saving and creating new images.
- Edit: editing and drawing in images.
- Image: modifying and converting images, geometric operations.

[8] http://rsb.info.nih.gov/ij/developer/macro/macros.html.

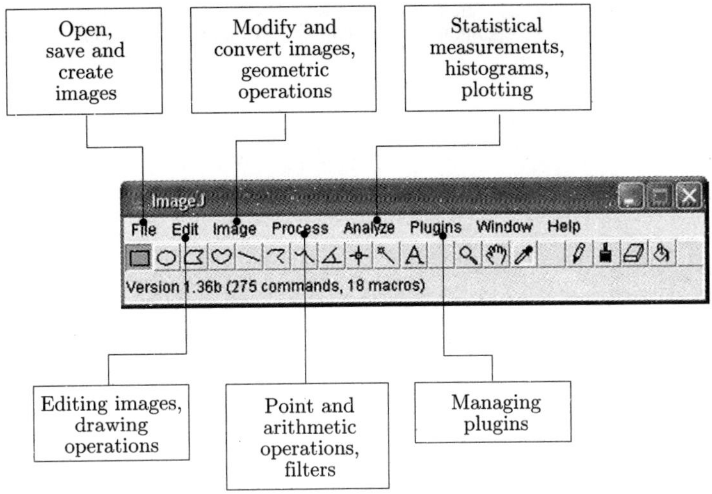

Figure 2.2 ImageJ main window (under Windows XP).

- Process: image processing, including point operations, filters, and arithmetic operations between multiple images.
- Analyze: statistical measurements on image data, histograms, and special display formats.
- Plugin: editing, compiling, executing, and managing user-defined plugins.

The current version of ImageJ can open images in several common formats, including TIFF (uncompressed only), JPEG, GIF, PNG, and BMP, as well as the formats DICOM[9] and FITS,[10] which are popular in medical and astronomical image processing, respectively. As is common in most image-editing programs, all interactive operations are applied to the currently *active* image, i. e., the image most recently selected by the user. ImageJ provides a simple (single-step) "undo" mechanism for most operations, which can also revert modifications effected by user-defined plugins.

2.2.3 ImageJ Plugins

Plugins are small Java modules for extending the functionality of ImageJ by using a simple standardized interface (Fig. 2.3). Plugins can be created, edited, compiled, invoked, and organized through the Plugin menu in ImageJ's main window (Fig. 2.2). Plugins can be grouped to improve modularity, and plugin

[9] Digital Imaging and Communications in Medicine.
[10] Flexible Image Transport System.

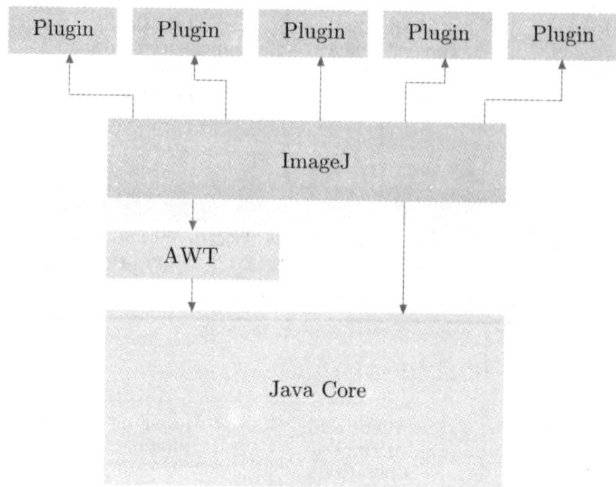

Figure 2.3 ImageJ software structure (simplified). ImageJ is based on the Java core system and depends in particular upon Java's Advanced Windowing Toolkit (AWT) for the implementation of the user interface and the presentation of image data. Plugins are small Java classes that extend the functionality of the basic ImageJ system.

commands can be arbitrarily placed inside the main menu structure. Also, many of ImageJ's built-in functions are actually implemented as plugins themselves.

Technically speaking, plugins are Java classes that implement a particular interface specification defined by ImageJ. There are two different kinds of plugins:

- `PlugIn`: requires no image to be open to start a plugin.
- `PlugInFilter`: the currently active image is passed to the plugin when started.

Throughout the examples in this book, we almost exclusively use plugins of the second type (`PlugInFilter`) for implementing image-processing operations. The interface specification requires that any plugin of type `PlugInFilter` must at least implement two methods, `setup()` and `run()`, with the following signatures:

`int setup (String` *arg*`, ImagePlus` *im*`)`
When the plugin is started, ImageJ calls this method first to verify that the capabilities of this plugin match the target image. `setup()` returns a vector of binary flags (packaged as a 32-bit `int` value) that describes the plugin's properties.

`void run (ImageProcessor ip)`
: This method does the actual work for this plugin. It is passed a single argument *ip*, an object of type `ImageProcessor`, which contains the image to be processed and all relevant information about it. The `run()` method returns no result value (`void`) but may modify the passed image and create new images.

2.2.4 A First Example: Inverting an Image

Let us look at a real example to quickly illustrate this mechanism. The task of our first plugin is to invert any 8-bit grayscale image to turn a positive image into a negative. As we shall see later, inverting the intensity of an image is a typical *point operation*, which is discussed in detail in Chapter 4. In ImageJ, 8-bit grayscale images have pixel values ranging from 0 (black) to 255 (white), and we assume that the width and height of the image are M and N, respectively. The operation is very simple: the value of each image pixel $I(u, v)$ is replaced by its inverted value,

$$I(u, v) \leftarrow 255 - I(u, v),$$

for all image coordinates (u, v), with $u = 0 \ldots M-1$ and $v = 0 \ldots N-1$.

The plugin class: `My_Inverter`

We decide to name our first plugin "`My_Inverter`", which is both the name of the Java class and the name of the source file that contains it (Prog. 2.1). The underscore character ("`_`") in the name causes ImageJ to recognize this class as a plugin and to insert it automatically into the menu list at startup. The Java source code in file `My_Inverter.java` contains a few `import` statements, followed by the definition of the class `My_Inverter`, which implements the `PlugInFilter` interface (because it will be applied to an existing image).

The `setup()` *method*

When a plugin of type `PlugInFilter` is executed, ImageJ first invokes its `setup()` method to obtain information about the plugin itself. In this example, `setup()` only returns the value `DOES_8G` (a static `int` constant specified by the `PlugInFilter` interface), indicating that this plugin can handle 8-bit grayscale images (Prog. 2.1, line 8). The parameters `arg` and `im` of the `setup()` method are not used in this case (see also Exercise 2.4).

```java
1 import ij.ImagePlus;
2 import ij.plugin.filter.PlugInFilter;
3 import ij.process.ImageProcessor;
4
5 public class My_Inverter implements PlugInFilter {
6
7   public int setup (String arg, ImagePlus im) {
8     return DOES_8G; // this plugin accepts 8-bit grayscale images
9   }
10
11   public void run (ImageProcessor ip) {
12     int w = ip.getWidth();
13     int h = ip.getHeight();
14
15     // iterate over all image coordinates
16     for (int u = 0; u < w; u++) {
17       for (int v = 0; v < h; v++) {
18         int p = ip.getPixel(u, v);
19         ip.putPixel(u, v, 255-p); // invert
20       }
21     }
22   }
23
24 } // end of class My_Inverter
```

Program 2.1 ImageJ plugin for inverting 8-bit grayscale images (file `My_Inverter.java`).

The `run()` *method*

As mentioned above, the `run()` method of a `PlugInFilter` plugin receives an object (`ip`) of type `ImageProcessor`, which contains the image to be processed and all relevant information about it. First, we use the `ImageProcessor` methods `getWidth()` and `getHeight()` to query the size of the image referenced by `ip` (lines 12–13). Then we use two nested `for` loops (with loop variables `u`, `v` for the horizontal and vertical coordinates, respectively) to iterate over all image pixels (lines 16–17). For reading and writing the pixel values, we use two additional methods of the class `ImageProcessor`:

`int getPixel (int` u`, int` v`)`
: Returns the pixel value at position (u, v) or zero if (u, v) is outside the image bounds.

`void putPixel (int` u`, int` v`, int` a`)`
: Sets the pixel value at position (u, v) to the new value a. Does nothing if (u, v) is outside the image bounds.

Details on these and other methods can be found in the ImageJ reference [5] (available online at the books support site).

If we are sure that no coordinates outside the image bounds are ever accessed (as in `My_Inverter` in Prog. 2.1) and the inserted pixel values are guaranteed not to exceed the image processor's range, we can use the slightly faster methods `get()` and `set()` in place of `getPixel()` and `putPixel()`, respectively. The most efficient way to process the image is to avoid read/write methods altogether and directly access the elements of the corresponding pixel array.[11]

Editing, compiling, and executing the plugin

The source code of our plugin should be stored in a file

`My_Inverter.java`

located within `<ij>/plugins/`[12] or an immediate subdirectory. New plugin files can be created with ImageJ's Plugins→New... menu. ImageJ even provides a built-in Java editor for writing plugins, which is available through the Plugins→Edit... menu but unfortunately is of little use for serious programming. A better alternative is to use a modern editor or a professional Java programming environment, such as Eclipse,[13] NetBeans,[14] or JBuilder,[15] all of which are freely available.

For compiling plugins (to Java bytecode), ImageJ comes with its own Java compiler as part of its runtime environment.[16] To compile and execute the new plugin, simply use the menu

Plugins→Compile and Run...

Compilation errors are displayed in a separate log window. Once the plugin is compiled, the corresponding `.class` file is automatically loaded and the plugin is applied to the currently active image. An error message is displayed if no images are open or if the current image cannot be handled by that plugin.

At startup, ImageJ automatically loads all correctly named plugins found in the `<ij>/plugins/` directory (or any immediate subdirectory) and installs them in its Plugins menu. These plugins can be executed immediately without any recompilation. References to plugins can also be placed manually with the

Plugins→Shortcuts→Install Plugin...

---

[11] See Sec. 7.6 of the ImageJ Short Reference [5].

[12] `<ij>` denotes ImageJ's installation directory, and `<ij>/plugins/` is the default plugins path, which can be set to any other directory.

[13] www.eclipse.org.

[14] www.netbeans.org.

[15] www.borland.com.

[16] Currently only for Windows; for MacOS and Linux, consult the ImageJ installation manual.

command at any other position in the ImageJ menu tree. Sequences of plugin calls and other ImageJ commands may be recorded as macro programs with Plugins→Macros→Record.

Displaying and "undoing" results

Our first plugin in Prog. 2.1 did not create a new image but "destructively" modified the target image. This is not always the case, but plugins can also create additional images or compute only statistics, without modifying the original image at all. It may be surprising, though, that our plugin contains no commands for displaying the modified image. This is done automatically by ImageJ whenever it can be assumed that the image passed to a plugin was modified.[17] In addition, ImageJ automatically makes a copy ("snapshot") of the image before passing it to the `run()` method of a `PlugInFilter`-type plugin. This feature makes it possible to restore the original image (with the Edit→Undo menu) after the plugin has finished without any explicit precautions in the plugin code.

2.3 Additional Information on ImageJ and Java

In the following chapters, we mostly use concrete plugins and Java code to describe algorithms and data structures. This not only makes these examples immediately applicable, but they should also help in acquiring additional skills for using ImageJ in a step-by-step fashion. To keep the text compact, we often describe only the `run()` method of a particular plugin and additional class and method definitions, if they are relevant in the given context. The complete source code for these examples can of course be downloaded from the book's supporting Website.[18]

2.3.1 Resources for ImageJ

The short reference in [5] contains an overview of ImageJ's main capabilities and a short description of its key classes, interfaces, and methods. The complete and most current API reference, including source code, tutorials, and many example plugins, can be found on the official ImageJ Website. Another great source for any serious plugin programming is the tutorial by Werner Bailer [3].

[17] No automatic redisplay occurs if the `NO_CHANGES` flag is set in the return value of the plugin's `setup()` method.

[18] www.imagingbook.com.

2.3.2 Programming with Java

While this book does not require extensive Java skills from its readers, some elementary knowledge is essential for understanding or extending the given examples. There is a huge and still-growing number of introductory textbooks on Java, such as [2, 11, 13] and many others. For readers with programming experience who have not worked with Java before, we particularly recommend some of the tutorials on Sun's Java Website.[19] Also, in Appendix B of this book, readers will find a small compilation of specific Java topics that cause frequent problems or programming errors.

2.4 Exercises

Exercise 2.1
Install the current version of ImageJ on your computer and make yourself familiar with the built-in functions (open, convert, edit, and save images).

Exercise 2.2
Write a new ImageJ plugin that reflects a grayscale image horizontally (or vertically) using `My_Inverter.java` (Prog. 2.1) as a template. Test your new plugin with appropriate images of different sizes (odd, even, extremely small) and inspect the results carefully.

Exercise 2.3
Create an ImageJ plugin for 8-bit grayscale images of arbitrary size that paints a white frame (with pixel value 255) 10 pixels wide *into* the image (without increasing its size). Make sure that this plugin also works for very small images.

Exercise 2.4
Write a new ImageJ plugin that shifts an 8-bit grayscale image horizontally and cyclically until the original state is reached again. To display the modified image after each shift, a reference to the corresponding `ImagePlus` object is required (`ImageProcessor` has no display methods). The `ImagePlus` object is only accessible to the plugin's `setup()` method, which is automatically called before the `run()` method. Modify the definition in Prog. 2.1 to keep a reference and to redraw the `ImagePlus` object as follows:

```
public class XY_plugin implements PlugInFilter {

  ImagePlus im;    // instance variable of this plugin object

  public int setup(String arg, ImagePlus im) {
    this.im = im;    // keep a reference to the image im
```

[19] http://java.sun.com/docs/books/tutorial/.

```
    return DOES_8G;
  }

  public void run(ImageProcessor ip) {
    ...  // use ip to modify the image
    im.updateAndDraw();  // use im to redisplay the image
    ...
  }

} // end of class XY_plugin
```

3
Histograms

Histograms are used to depict image statistics in an easily interpreted visual format. With a histogram, it is easy to determine certain types of problems in an image, for example, it is simple to conclude if an image is properly exposed by visual inspection of its histogram. In fact, histograms are so useful that modern digital cameras often provide a real-time histogram overlay on the viewfinder (Fig. 3.1) to help prevent taking poorly exposed pictures. It is important to catch errors like this at the image capture stage because poor exposure results in a permanent loss of information which it is not possible to recover later using image-processing techniques. In addition to their usefulness during image capture, histograms are also used later to improve the visual appearance of an image and as a "forensic" tool for determining what type of processing has previously been applied to an image.

3.1 What Is a Histogram?

Histograms in general are frequency distributions, and histograms of images describe the frequency of the intensity values that occur in an image. This concept can be easily explained by considering an old-fashioned grayscale image like the one shown in Fig. 3.2. A histogram h for a grayscale image I with intensity values in the range $I(u,v) \in [0, K-1]$ would contain exactly K entries, where for a typical 8 bit grayscale image, $K = 2^8 = 256$. Each individual histogram entry is defined as

$$\mathsf{h}(i) = \text{ the } \textit{number} \text{ of pixels in } I \text{ with the intensity value } i,$$

W. Burger, M.J. Burge, *Principles of Digital Image Processing*, Undergraduate Topics in Computer Science, DOI 10.1007/978-1-84800-191-6_3, © Springer-Verlag London Limited, 2009

Figure 3.1 Digital camera back display showing a histogram overlay.

Figure 3.2 An 8-bit grayscale image and a histogram depicting the frequency distribution of its 256 intensity values.

for all $0 \leq i < K$. More formally stated,

$$\mathsf{h}(i) = \mathrm{card}\big\{(u,v) \mid I(u,v) = i\,\big\}.^{1} \tag{3.1}$$

Therefore $\mathsf{h}(0)$ is the number of pixels with the value 0, $\mathsf{h}(1)$ the number of pixels with the value 1, and so forth. Finally $\mathsf{h}(255)$ is the number of all white pixels with the maximum intensity value $255 = K-1$. The result of the histogram computation is a one-dimensional vector h of length K. Figure 3.3 gives an example for an image with $K = 16$ possible intensity values.

Since a histogram encodes no information about *where* each of its individual entries originated in the image, histograms contain no information about the spatial arrangement of pixels in the image. This is intentional since the

[1] card{...} denotes the number of elements ("cardinality") in a set (see also p. 233).

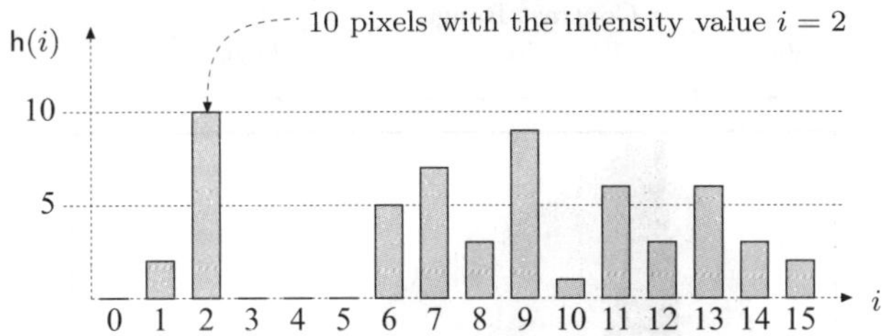

h(i)	0	2	10	0	0	0	5	7	3	9	1	6	3	6	3	2
i	0	1	2	3	4	5	6	7	8	9	10	11	12	13	14	15

Figure 3.3 Histogram vector for an image with $K = 16$ possible intensity values. The indices of the vector element $i = 0 \ldots 15$ represent intensity values. The value of 10 at index 2 means that the image contains 10 pixels of intensity value 2.

Figure 3.4 Three very different images with identical histograms.

main function of a histogram is to provide statistical information, (e.g., the distribution of intensity values) in a compact form. Is it possible to reconstruct an image using only its histogram? That is, can a histogram be somehow "inverted"? Given the loss of spatial information, in all but the most trivial cases, the answer is no. As an example, consider the wide variety of images you could construct using the same number of pixels of a specific value. These images would appear different but have exactly the same histogram (Fig. 3.4).

3.2 Interpreting Histograms

A histogram depicts problems that originate during image acquisition, such as those involving contrast and dynamic range, as well as artifacts resulting from image-processing steps that were applied to the image. Histograms are often used to determine if an image is making effective use of its intensity range (Fig. 3.5) by examining the size and uniformity of the histogram's distribution.

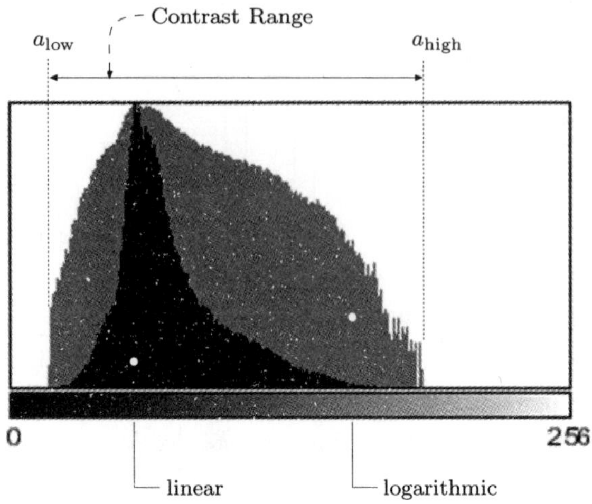

Figure 3.5 The effective intensity range. The graph depicts how often a pixel value occurs linearly (black bars) and logarithmically (gray bars). The logarithmic form makes even relatively low occurrences, which can be very important in the image, readily apparent.

3.2.1 Image Acquisition

Exposure

Histograms make classic exposure problems readily apparent. As an example, a histogram where a large span of the intensity range at one end is largely unused while the other end is crowded with high-value peaks (Fig. 3.6) is representative of an improperly exposed image.

Figure 3.6 Exposure errors are readily apparent in histograms. Underexposed (a), properly exposed (b), and overexposed (c) photographs.

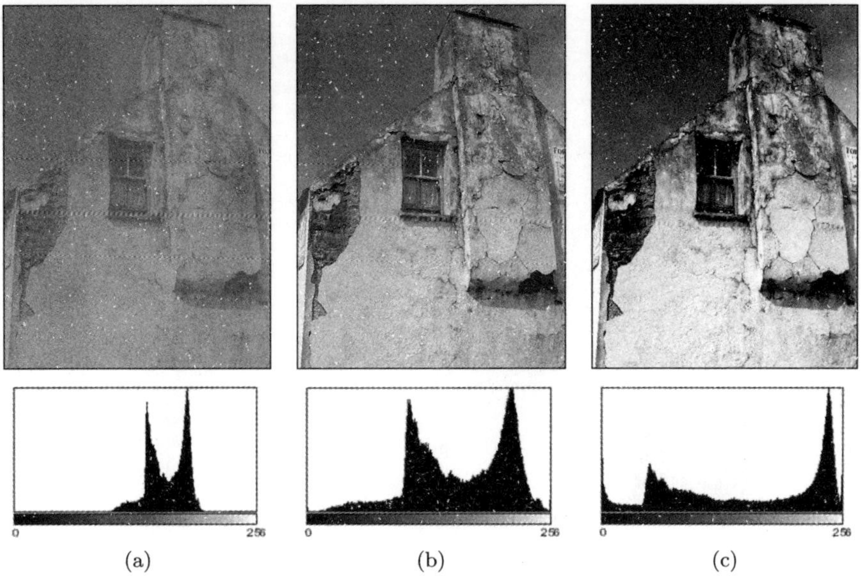

Figure 3.7 How changes in contrast affect a histogram: low contrast (a), normal contrast (b), high contrast (c).

Contrast

Contrast is understood as the range of intensity values *effectively* used within a given image, that is the difference between the image's maximum and minimum pixel values. A full-contrast image makes effective use of the entire range of available intensity values from $a = a_{\min} \ldots a_{\max} = 0 \ldots K-1$ (black to white). Using this definition, image contrast can be easily read directly from the histogram. Figure 3.7 illustrates how varying the contrast of an image affects its histogram.

Dynamic range

The dynamic range of an image is, in principle, understood as the number of *distinct* pixel values in an image. In the ideal case, the dynamic range encompasses all K usable pixel values, in which case the value range is completely utilized. When an image has an available range of contrast $a = a_{\mathrm{low}} \ldots a_{\mathrm{high}}$, with

$$a_{\min} < a_{\mathrm{low}} \qquad \text{and} \qquad a_{\mathrm{high}} < a_{\max},$$

then the maximum possible dynamic range is achieved when all the intensity values lying in this range are utilized (i. e., appear in the image; Fig. 3.8).

While the contrast of an image can be increased by transforming its existing values so that they utilize more of the underlying value range available, the dy-

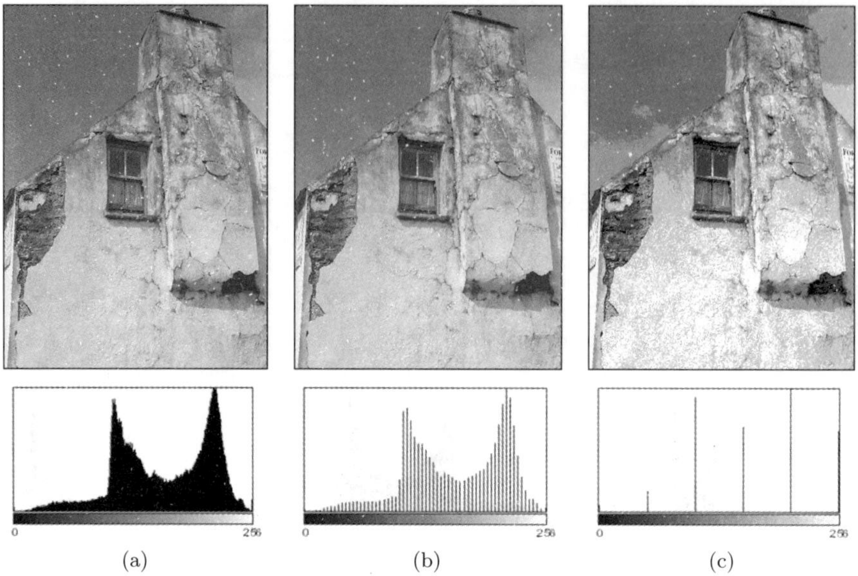

Figure 3.8 How changes in dynamic range affect a histogram: high dynamic range (a), low dynamic range with 64 intensity values (b), extremely low dynamic range with only 6 intensity values (c).

namic range of an image can only be increased by introducing artificial (that is, not originating with the image sensor) values using methods such as interpolation (see Vol. 2 [6, Sec. 10.3]). An image with a high dynamic range is desirable because it will suffer less image-quality degradation during image processing and compression. Since it is not possible to increase dynamic range after image acquisition in a practical way, professional cameras and scanners work at depths of more than 8 bits, often 12–14 bits per channel, in order to provide high dynamic range at the acquisition stage. While most output devices, such as monitors and printers, are unable to actually reproduce more than 256 different shades, a high dynamic range is always beneficial for subsequent image processing or archiving.

3.2.2 Image Defects

Histograms can be used to detect a wide range of image defects that originate either during image acquisition or as the result of later image processing. Since histograms always depend on the visual characteristics of the scene captured in the image, no single "ideal" histogram exists. While a given histogram may be optimal for a specific scene, it may be entirely unacceptable for another. As an example, the ideal histogram for an astronomical image would likely be very different from that of a good landscape or portrait photo. Nevertheless, there are some general rules; for example, when taking a landscape image with a digital camera, you can expect the histogram to have evenly distributed intensity values and no isolated spikes.

Saturation

Ideally the contrast range of a sensor, such as that used in a camera, should be greater than the range of the intensity of the light that it receives from a scene. In such a case, the resulting histogram will be smooth at both ends because the light received from the very bright and the very dark parts of the scene will be less than the light received from the other parts of the scene. Unfortunately, this ideal is often not the case in reality, and illumination outside of the sensor's contrast range, arising for example from glossy highlights and especially dark parts of the scene, cannot be captured and is lost. The result is a histogram that is saturated at one or both ends of its range. The illumination values lying outside of the sensor's range are mapped to its minimum or maximum values and appear on the histogram as significant spikes at the tail ends. This typically occurs in an under- or overexposed image and is generally not avoidable when the inherent contrast range of the scene exceeds the range of the system's sensor (Fig. 3.9 (a)).

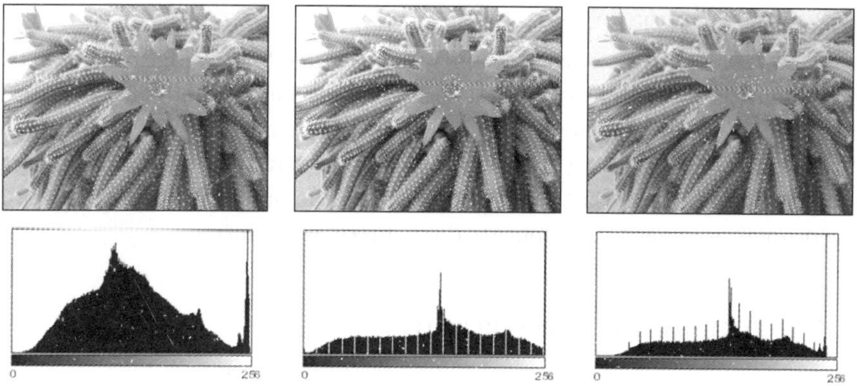

(a) (b) (c)

Figure 3.9 Effect of image capture errors on histograms: saturation of high intensities (a), histogram gaps caused by a slight increase in contrast (b), and histogram spikes resulting from a reduction in contrast (c).

Spikes and gaps

As discussed above, the intensity value distribution for an unprocessed image is generally smooth; that is, it is unlikely that isolated spikes (except for possible saturation effects at the tails) or gaps will appear in its histogram. It is also unlikely that the count of any given intensity value will differ greatly from that of its neighbors (i. e., it is locally smooth). While artifacts like these are observed very rarely in original images, they will often be present after an image has been manipulated, for instance, by changing its contrast. Increasing the contrast (see Ch. 4) causes the histogram lines to separate from each other and, due to the discrete values, gaps are created in the histogram (Fig. 3.9 (b)). Decreasing the contrast leads, again because of the discrete values, to the merging of values that were previously distinct. This results in increases in the corresponding histogram entries and ultimately leads to highly visible spikes in the histogram (Fig. 3.9 (c)).[2]

Impacts of image compression

Image compression also changes an image in ways that are immediately evident in its histogram. As an example, during GIF compression, an image's dynamic range is reduced to only a few intensities or colors, resulting in an obvious line structure in the histogram that cannot be removed by subsequent processing (Fig. 3.10). Generally, a histogram can quickly reveal whether an image has ever been subjected to color quantization, such as occurs during conversion to a GIF image, even if the image has subsequently been converted to a full-color format such as TIFF or JPEG.

Figure 3.11 illustrates what occurs when a simple line graphic with only two gray values (128, 255) is subjected to a compression method such as JPEG, that is not designed for line graphics but instead for natural photographs. The histogram of the resulting image clearly shows that it now contains a large number of gray values that were not present in the original image, resulting in a poor-quality image[3] that appears dirty, fuzzy, and blurred.

3.3 Computing Histograms

Computing the histogram of an 8-bit grayscale image containing intensity values between 0 and 255 is a simple task. All we need is a set of 256 counters, one for each possible intensity value. First, all counters are initialized to zero.

---

[2] Unfortunately, these types of errors are also caused by the internal contrast "optimization" routines of some image-capture devices, especially consumer-type scanners.

[3] Using JPEG compression on images like this, for which it was not designed, is one of the most egregious of imaging errors. JPEG is designed for photographs of natural scenes with smooth color transitions, and using it to compress iconic images with large areas of the same color results in strong visual artifacts (see, for example, Fig. 1.9 on p. 19).

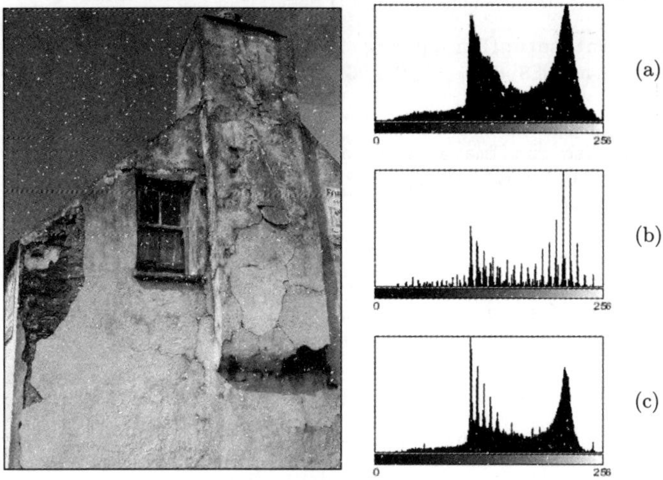

Figure 3.10 Color quantization effects resulting from GIF conversion. The original image converted to a 256 color GIF image (left). Original histogram (a) and the histogram after GIF conversion (b). When the RGB image is scaled by 50%, some of the lost colors are recreated by interpolation, but the results of the GIF conversion remain clearly visible in the histogram (c).

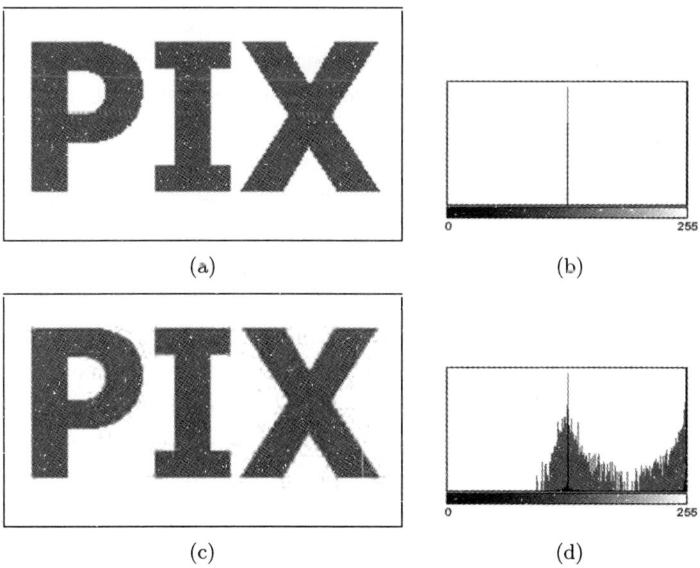

Figure 3.11 Effects of JPEG compression. The original image (a) contained only two different gray values, as its histogram (b) makes readily apparent. JPEG compression, a poor choice for this type of image, results in numerous additional gray values, which are visible in both the resulting image (c) and its histogram (d). In both histograms, the linear frequency (black bars) and the logarithmic frequency (gray bars) are shown.

```
1  public class Compute_Histogram implements PlugInFilter {
2
3      public int setup(String arg, ImagePlus img) {
4          return DOES_8G + NO_CHANGES;
5      }
6
7      public void run(ImageProcessor ip) {
8          int[] H = new int[256]; // histogram array
9          int w = ip.getWidth();
10         int h = ip.getHeight();
11
12         for (int v = 0; v < h; v++) {
13             for (int u = 0; u < w; u++) {
14                 int i = ip.getPixel(u,v);
15                 H[i] = H[i] + 1;
16             }
17         }
18         ... //histogram H[] can now be used
19     }
20
21 } // end of class Compute_Histogram
```

Program 3.1 ImageJ plugin for computing the histogram of an 8-bit grayscale image. The `setup()` method returns `DOES_8G + NO_CHANGES`, which indicates that this plugin requires an 8-bit grayscale image and will not alter it (line 4). In Java, all elements of a newly instantiated array (line 8) are automatically initialized, in this case to zero.

Then we iterate through the image I, determining the pixel value p at each location (u, v), and incrementing the corresponding counter by one. At the end, each counter will contain the number of pixels in the image that have the corresponding intensity value.

An image with K possible intensity values requires exactly K counter variables; for example, since an 8-bit grayscale image can contain at most 256 different intensity values, we require 256 counters. While individual counters make sense conceptually, an actual implementation would not use K individual *variables* to represent the counters but instead would use an *array* with K entries (`int[256]` in Java). In this example, the actual implementation as an array is straightforward. Since the intensity values begin at zero (like arrays in Java) and are all positive, they can be used directly as the indices $i \in [0, N-1]$ of the histogram array. Program 3.1 contains the complete Java source code for computing a histogram within the `run()` method of an ImageJ plugin.

At the start of Prog. 3.1, the array `H` of type `int[]` is created (line 8) and its elements are automatically initialized[4] to `0`. It makes no difference, at least in terms of the final result, whether the array is traversed in row or column

[4] In Java, arrays of primitives such as `int`, `double` are initialized at creation to `0` in the case of integer types or `0.0` for floating-point types, while arrays of objects are initialized to `null`.

order, as long as all pixels in the image are visited exactly once. In contrast to Prog. 2.1, in this example we traverse the array in the standard row-first order such that the outer `for` loop iterates over the *vertical* coordinates v and the inner loop over the *horizontal* coordinates u.[5] Once the histogram has been calculated, it is available for further processing steps or for being displayed.

Of course, histogram computation is already implemented in ImageJ and is available via the method `getHistogram()` for objects of the class `ImageProcessor`. If we use this built-in method, the `run()` method of Prog. 3.1 can be simplified to

```
public void run(ImageProcessor ip) {
    int[] H = ip.getHistogram();  // built-in ImageJ method
    ... // histogram H[] can now be used
}
```

3.4 Histograms of Images with More than 8 Bits

Normally histograms are computed in order to visualize the image's distribution on the screen. This presents no problem when dealing with images having $2^8 = 256$ entries, but when an image uses a larger range of values, for instance 16- and 32-bit or floating-point images (see Table 1.1), then the growing number of necessary histogram entries makes this no longer practical.

3.4.1 Binning

Since it is not possible to represent each intensity value with its own entry in the histogram, we will instead let a given entry in the histogram represent a *range* of intensity values. This technique is often referred to as "binning" since you can visualize it as collecting a range of pixel values in a container such as a bin or bucket. In a binned histogram of size B, each bin $\mathsf{h}(j)$ contains the number of image elements having values within the interval $a_j \leq a < a_{j+1}$, and therefore (analogous to Eqn. (3.1))

$$\mathsf{h}(j) = \text{card}\left\{(u,v) \mid a_j \leq I(u,v) < a_{j+1}\right\}, \quad \text{for } 0 \leq j < B. \tag{3.2}$$

Typically the range of possible values in B is divided into bins of equal size $k_B = K/B$ such that the starting value of the interval j is

$$a_j = j \cdot \frac{K}{B} = j \cdot k_B.$$

[5] In this way, image elements are traversed in exactly the same way that they are laid out in computer memory, resulting in more efficient memory access and with it the possibility of increased performance, especially when dealing with larger images (see also Appendix B, p. 242).

3.4.2 Example

In order to create a typical histogram containing $B = 256$ entries from a 14-bit image, you would divide the available value range if $j = 0 \ldots 2^{14}-1$ into 256 equal intervals, each of length $k_B = 2^{14}/256 = 64$, so that $a_0 = 0$, $a_1 = 64$, $a_2 = 128$, ... $a_{255} = 16{,}320$ and $a_{256} = a_B = 2^{14} = 16{,}384 = K$. This results in the following mapping from the pixel values to the histogram bins $\mathsf{h}(0) \ldots \mathsf{h}(255)$:

$$
\begin{array}{lcrcl}
\mathsf{h}(0) & \leftarrow & 0 \leq I(u,v) < & 64 \\
\mathsf{h}(1) & \leftarrow & 64 \leq I(u,v) < & 128 \\
\mathsf{h}(2) & \leftarrow & 128 \leq I(u,v) < & 192 \\
\vdots & & \vdots \quad \vdots & \vdots \\
\mathsf{h}(j) & \leftarrow & a_j \leq I(u,v) < & a_{j+1} \\
\vdots & & \vdots \quad \vdots & \vdots \\
\mathsf{h}(255) & \leftarrow & 16320 \leq I(u,v) < & 16384
\end{array}
$$

3.4.3 Implementation

If, as in the above example, the value range $0 \ldots K-1$ is divided into equal length intervals $k_B = K/B$, there is naturally no need to use a mapping table to find a_j since for a given pixel value $a = I(u,v)$ the correct histogram element j is easily computed. In this case, it is enough to simply divide the pixel value $I(u,v)$ by the interval length k_B; that is,

$$\frac{I(u,v)}{k_B} = \frac{I(u,v)}{K/B} = \frac{I(u,v) \cdot B}{K} . \tag{3.3}$$

As an index to the appropriate histogram bin $\mathsf{h}(j)$, we require an integer value

$$j = \left\lfloor \frac{I(u,v) \cdot B}{K} \right\rfloor, \tag{3.4}$$

where $\lfloor \cdot \rfloor$ denotes the *floor* function.[6] A Java method for computing histograms by "linear binning" is given in Prog. 3.2. Note that all the computations from Eqn. (3.4) are done with integer numbers without using any floating-point operations. Also there is no need to explicitly call the *floor* function because the expression

```
a * B / K
```

in line 11 uses integer division and in Java the fractional result of such an operation is truncated, which is equivalent to applying the floor function (assuming

```
int[] binnedHistogram(ImageProcessor ip) {
  int K = 256; // number of intensity values
  int B = 32; // size of histogram, must be defined
  int[] H = new int[B]; // histogram array
  int w = ip.getWidth();
  int h = ip.getHeight();

  for (int v = 0; v < h; v++) {
    for (int u = 0; u < w; u++) {
      int a = ip.getPixel(u, v);
      int i = a * B / K; // integer operations only!
      H[i] = H[i] + 1;
    }
  }
  // return binned histogram
  return H;
}
```

Program 3.2 Histogram computation using "binning" (Java method). Example of computing a histogram with $B = 32$ bins for an 8-bit grayscale image with $K = 256$ intensity levels. The method `binnedHistogram()` returns the histogram of the image object `ip` passed to it as an `int` array of size B.

positive arguments).[7] The binning method can also be applied, in a similar way, to floating-point images.

3.5 Color Image Histograms

When referring to histograms of color images, typically what is meant is a histogram of the image intensity (luminance) or of the individual color channels. Both of these variants are supported by practically every image-processing application and are used to objectively appraise the image quality, especially directly after image acquisition.

3.5.1 Intensity Histograms

The intensity or *luminance* histogram $\mathsf{h}_{\mathrm{Lum}}$ of a color image is nothing more than the histogram of the corresponding grayscale image, so naturally all aspects of the preceding discussion also apply to this type of histogram. The grayscale image is obtained by computing the luminance of the individual channels of the color image. When computing the luminance, it is not sufficient to simply average the values of each color channel; instead, a weighted sum that

[6] $\lfloor x \rfloor$ rounds x down to the next whole number (see Appendix A, p. 233).

[7] For a more detailed discussion, see the section on integer division in Java in Appendix B (p. 237).

takes into account color perception theory should be computed. This process is explained in detail in Chapter 8 (p. 202).

3.5.2 Individual Color Channel Histograms

Even though the luminance histogram takes into account all color channels, image errors appearing in single channels can remain undiscovered. For example, the luminance histogram may appear clean even when one of the color channels is oversaturated. In RGB images, the blue channel contributes only a small amount to the total brightness and so is especially sensitive to this problem.

Component histograms supply additional information about the intensity distribution within the individual color channels. When computing component histograms, each color channel is considered a separate intensity image and each histogram is computed independently of the other channels. Figure 3.12 shows the luminance histogram $\mathsf{h}_{\mathrm{Lum}}$ and the three component histograms h_{R}, h_{G}, and h_{B} of a typical RGB color image. Notice that saturation problems in all three channels (red in the upper intensity region, green and blue in the lower regions) are obvious in the component histograms but not in the luminance histogram. In this case it is striking, and not at all atypical, that the three component histograms appear completely different from the corresponding luminance histogram $\mathsf{h}_{\mathrm{Lum}}$ (Fig. 3.12 (b)).

3.5.3 Combined Color Histograms

Luminance histograms and component histograms both provide useful information about the lighting, contrast, dynamic range, and saturation effects relative to the individual color components. It is important to remember that they provide no information about the distribution of the actual *colors* in the image because they are based on the individual color channels and not the combination of the individual channels that forms the color of an individual pixel. Consider, for example, when h_{R}, the component histogram for the red channel, contains the entry

$$\mathsf{h}_{\mathrm{R}}(200) = 24.$$

Then it is only known that the image has 24 pixels that have a red intensity value of 200. The entry does not tell us anything about the green and blue values of those pixels, which could be any valid value $(*)$; that is,

$$(r, g, b) = (200, *, *).$$

Suppose further that the three component histograms included the following entries:

$$\mathsf{h}_{\mathrm{R}}(50) = 100, \quad \mathsf{h}_{\mathrm{G}}(50) = 100, \quad \mathsf{h}_{\mathrm{B}}(50) = 100.$$

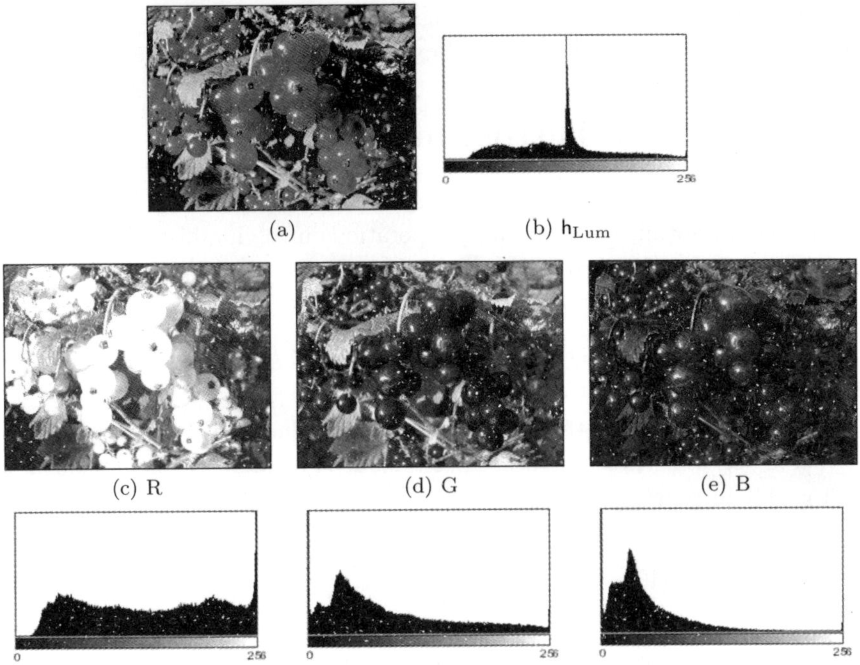

(a) (b) $\mathsf{h}_{\mathrm{Lum}}$

(c) R (d) G (e) B

(f) h_{R} (g) h_{G} (h) h_{B}

Figure 3.12 Histograms of an RGB color image: original image (a), luminance histogram $\mathsf{h}_{\mathrm{Lum}}$ (b), RGB color components as intensity images (c–e), and the associated component histograms h_{R}, h_{G}, h_{B} (f–h). The fact that all three color channels have saturation problems is only apparent in the individual component histograms. The spike in the distribution resulting from this is found in the middle of the luminance histogram (b).

Could we conclude from this that the image contains 100 pixels with the color combination

$$(r, g, b) = (50, 50, 50)$$

or that this color occurs at all? In general, no, because there is no way of ascertaining from these data if there exists a pixel in the image in which all three components have the value 50. The only thing we could really say is that the color value $(50, 50, 50)$ can occur at most 100 times in this image.

So, although conventional (intensity or component) histograms of color images depict important properties, they do not really provide any useful information about the composition of the actual colors in an image. In fact, a collection of color images can have very similar component histograms and still contain entirely different colors. This leads to the interesting topic of the *combined* histogram, which uses statistical information about the combined color components in an attempt to determine if two images are roughly similar in their color composition. Features computed from this type of histogram often form the foundation of color-based image retrieval methods. We will return to this topic in Chapter 8, where we will explore color images in greater detail.

3.6 Cumulative Histogram

The cumulative histogram, which is derived from the ordinary histogram, is useful when performing certain image operations involving histograms; for instance, histogram equalization (see Sec. 4.5). The cumulative histogram H is defined as

$$\mathsf{H}(i) = \sum_{j=0}^{i} \mathsf{h}(j) \quad \text{for } 0 \leq i < K. \tag{3.5}$$

A particular value $\mathsf{H}(i)$ is thus the sum of all the values $\mathsf{h}(j)$, with $j \leq i$, in the original histogram. Alternatively, we can define H recursively (as implemented in Prog. 4.2 on p. 66):

$$\mathsf{H}(i) = \begin{cases} \mathsf{h}(0) & \text{for } i = 0 \\ \mathsf{H}(i-1) + \mathsf{h}(i) & \text{for } 0 < i < K. \end{cases} \tag{3.6}$$

The cumulative histogram $\mathsf{H}(i)$ is a monotonically increasing function with a maximum value

$$\mathsf{H}(K-1) = \sum_{j=0}^{K-1} \mathsf{h}(j) = M \cdot N; \tag{3.7}$$

that is, the total number of pixels in an image of width M and height N. Figure 3.13 shows a concrete example of a cumulative histogram.

The cumulative histogram is useful not primarily for viewing but as a simple and powerful tool for capturing statistical information from an image. In particular, we will use it in the next chapter to compute the parameters for several common point operations (see Sections 4.4–4.6).

3.7 Exercises

Exercise 3.1

In Prog. 3.2, `B` and `K` are constants. Consider if there would be an advantage to computing the value of `B/K` outside of the loop, and explain your reasoning.

Exercise 3.2

Develop an ImageJ plugin that computes the cumulative histogram of an 8-bit grayscale image and displays it as a new image, similar to $\mathsf{H}(i)$ in Fig. 3.13.

Hint: Use the `ImageProcessor` method `int[] getHistogram()` to retrieve

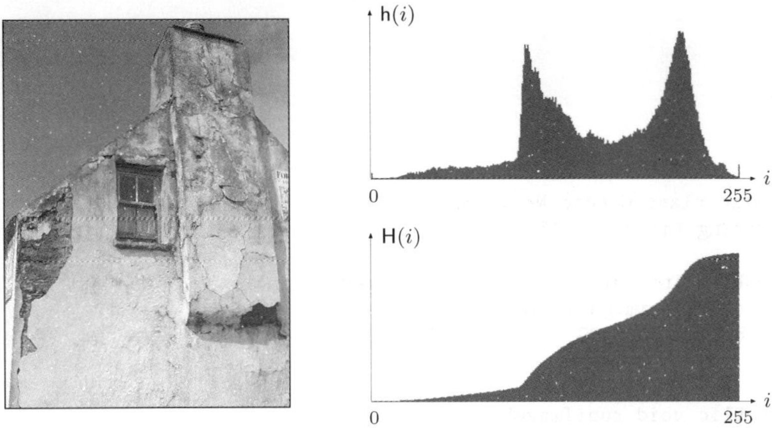

Figure 3.13 The ordinary histogram $\mathsf{h}(i)$ and its associated cumulative histogram $\mathsf{H}(i)$.

the original image's histogram values and then compute the cumulative histogram "in place" according to Eqn. (3.6). Create a new (blank) image of appropriate size (e. g., 256×150) and draw the scaled histogram data as black vertical bars such that the maximum entry spans the full height of the image. Program 3.3 shows how this plugin could be set up and how a new image is created and displayed.

Exercise 3.3

Develop a technique for nonlinear binning that uses a table of interval limits a_j (Eqn. (3.2)).

Exercise 3.4

Develop an ImageJ plugin that uses the Java methods `Math.random()` or `Random.nextInt(int n)` to create an image with random pixel values that are uniformly distributed in the range $[0, 255]$. Analyze the image's histogram to determine how equally distributed the pixel values truly are.

Exercise 3.5

Develop an ImageJ plugin that creates a random image with a Gaussian (normal) distribution with mean value $\mu = 128$ and standard deviation $\sigma = 50$. Use the standard Java method `double Random.nextGaussian()` to produce normally-distributed random numbers (with $\mu = 0$ and $\sigma = 1$) and scale them appropriately to pixel values. Analyze the resulting image histogram to see if it shows a Gaussian distribution too.

```
1 public class Create_New_Image implements PlugInFilter {
2   String title = null;
3
4   public int setup(String arg, ImagePlus im) {
5     title = im.getTitle();
6     return DOES_8G + NO_CHANGES;
7   }
8
9   public void run(ImageProcessor ip) {
10    int w = 256;
11    int h = 100;
12    int[] hist = ip.getHistogram();
13
14    // create the histogram image:
15    ImageProcessor histIp = new ByteProcessor(w, h);
16    histIp.setValue(255); // white = 255
17    histIp.fill(); // clear this image
18
19    // draw the histogram values as black bars in ip2 here,
20    // for example, using histIp.putpixel(u,v,0)
21    // ...
22
23    // display the histogram image:
24    String hTitle = "Histogram of " + title;
25    ImagePlus histIm = new ImagePlus(hTitle, histIp);
26    histIm.show();
27    // histIm.updateAndDraw();
28  }
29
30 } // end of class Create_New_Image
```

Program 3.3 Creating and displaying a new image (ImageJ plugin). First, we create a `ByteProcessor` object (`histIp`, line 15) that is subsequently filled. At this point, `histIp` has no screen representation and is thus not visible. Then, an associated `ImagePlus` object is created (line 25) and displayed by applying the `show()` method (line 26). Notice how the title (`String`) is retrieved from the original image inside the `setup()` method (line 5) and used to compose the new image's title (lines 24 and 25). If `histIp` is changed *after* calling `show()`, then the method `updateAndDraw()` could be used to redisplay the associated image again (line 27).

4 Point Operations

Point operations perform a modification of the pixel values without changing the size, geometry, or local structure of the image. Each new pixel value $a' = I'(u,v)$ depends exclusively on the previous value $a = I(u,v)$ at the *same* position and is thus independent from any other pixel value, in particular from any of its neighboring pixels.[1] The original pixel values are mapped to the new values by a function $f(a)$,

$$\begin{aligned} a' &\leftarrow f(a) \qquad \text{or} \\ I'(u,v) &\leftarrow f\big(I(u,v)\big), \end{aligned} \tag{4.1}$$

for each image position (u,v). If the function $f()$ is independent of the image coordinates (i. e., the same throughout the image), the operation is called "global" or "homogeneous". Typical examples of homogeneous point operations include, among others,

- modifying image brightness or contrast,
- applying arbitrary intensity transformations ("curves"),
- quantizing (or "posterizing") images,
- global thresholding,
- gamma correction,
- color transformations.

[1] If the result depends on more than one pixel value, the operation is called a "filter", as described in Ch. 5.

W. Burger, M.J. Burge, *Principles of Digital Image Processing*, Undergraduate Topics in Computer Science, DOI 10.1007/978-1-84800-191-6_4, © Springer-Verlag London Limited, 2009

We will look at some of these techniques in more detail in the following.

In contrast, the mapping function $g()$ for a *nonhomogeneous* point operation would also take into account the current image coordinate (u, v); i. e.,

$$
\begin{aligned}
a' &\leftarrow g(a, u, v) \qquad \text{or} \\
I'(u, v) &\leftarrow g\left(I(u, v), u, v\right).
\end{aligned}
\tag{4.2}
$$

A typical nonhomogeneous operation is the local adjustment of contrast or brightness used for example to compensate for uneven lighting during image acquisition.

4.1 Modifying Image Intensity

4.1.1 Contrast and Brightness

Let us start with a simple example. Increasing the image's contrast by 50% (i. e., by the factor 1.5) or raising the brightness by 10 units can be expressed by the mapping functions

$$f_{\mathrm{contr}}(a) = a \cdot 1.5 \qquad \text{and} \qquad f_{\mathrm{bright}}(a) = a + 10, \tag{4.3}$$

respectively. The first operation is implemented as an ImageJ plugin by the code shown in Prog. 4.1, which can easily be adapted to perform any other type of point operation. Rounding to the nearest integer values is accomplished by simply adding 0.5 before the truncation effected by the `(int)` typecast in line 7 (this only works for positive values). Also note the use of the more efficient image processor methods `get()` and `set()` (instead of `getPixel()` and `putPixel()`) in this example.

4.1.2 Limiting the Results by Clamping

When implementing arithmetic operations on pixels, we must keep in mind that the computed results may exceed the maximum range of pixel values for a given image type ($[0 \ldots 255]$ in the case of 8-bit grayscale images). To avoid this, we have included the "clamping" statement

```
if (a > 255)  a = 255;
```

in line 9 of Prog. 4.1, which limits any result to the maximum value 255. Similarly one should, in general, also limit the results to the minimum value (0) to avoid negative pixel values (which cannot be represented by this type of 8-bit image), for example by the statement

```
if (a < 0)  a = 0;
```

This second measure is not necessary in Prog. 4.1 because the intermediate results can never be negative in this particular operation.

```
1  public void run(ImageProcessor ip) {
2    int w = ip.getWidth();
3    int h = ip.getHeight();
4
5    for (int v = 0; v < h; v++) {
6      for (int u = 0; u < w; u++) {
7        int a = (int) (ip.get(u, v) * 1.5 + 0.5);
8        if (a > 255)
9          a = 255;     // clamp to maximum value
10       ip.set(u, v, a);
11     }
12   }
13 }
```

Program 4.1 Point operation to increase the contrast by 50% (ImageJ plugin). Note that in line 7 the result of the multiplication of the integer pixel value by the constant 1.5 (implicitly of type `double`) is of type `double`. Thus an explicit type cast `(int)` is required to assign the value to the `int` variable `a`. 0.5 is added in line 7 to round to the nearest integer values.

4.1.3 Inverting Images

Inverting an intensity image is a simple point operation that reverses the ordering of pixel values (by multiplying with -1) and adds a constant value to map the result to the admissible range again. Thus, for a pixel value $a = I(u, v)$ in the range $[0, a_{\max}]$, the corresponding point operation is

$$f_{\text{invert}}(a) = -a + a_{\max} = a_{\max} - a. \tag{4.4}$$

The inversion of an 8-bit grayscale image with $a_{\max} = 255$ was the task of our first plugin example in Sec. 2.2.4 (Prog. 2.1). Note that in this case no clamping is required at all because the function always maps to the original range of values. In ImageJ, this operation is performed by the method `invert()` (for objects of type `ImageProcessor`) and is also available through the Edit→Invert menu. Obviously, inverting an image mirrors its histogram, as shown in Fig. 4.5 (c).

4.1.4 Threshold Operation

Thresholding an image is a special type of quantization that separates the pixel values in two classes, depending upon a given threshold value a_{th} that is usually constant. The threshold function $f_{\text{threshold}}(a)$ maps all pixels to one of two fixed intensity values a_0 or a_1; i. e.,

$$f_{\text{threshold}}(a) = \begin{cases} a_0 & \text{for } a < a_{\text{th}} \\ a_1 & \text{for } a \geq a_{\text{th}} \end{cases} \tag{4.5}$$

with $0 < a_{\text{th}} \leq a_{\max}$. A common application is *binarizing* an intensity image with the values $a_0 = 0$ and $a_1 = 1$.

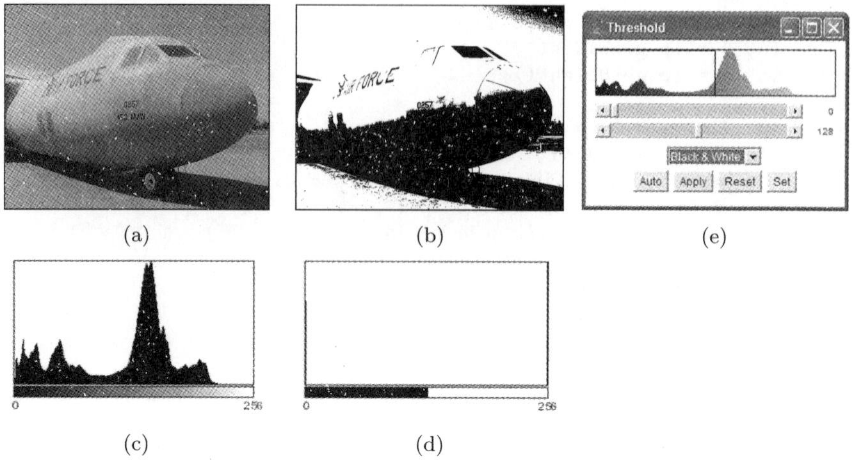

Figure 4.1 Threshold operation: original image (a) and corresponding histogram (c); result after thresholding with $a_{\mathrm{th}} = 128$, $a_0 = 0$, $a_1 = 255$ (b) and corresponding histogram (d); ImageJ's interactive Threshold menu (e).

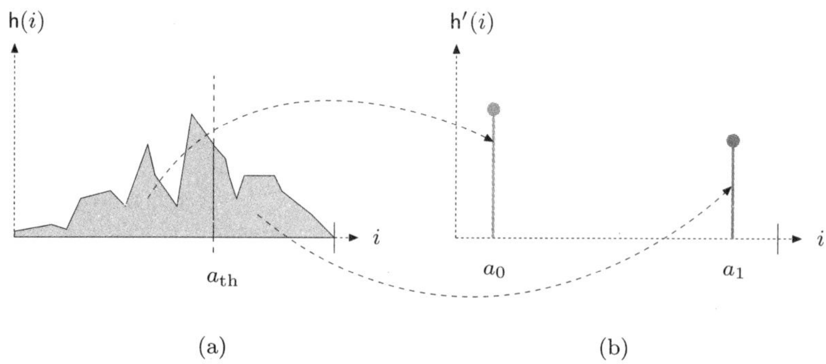

Figure 4.2 Effects of thresholding upon the histogram. The threshold value is a_{th}. The original distribution (a) is split and merged into two isolated entries at a_0 and a_1 in the resulting histogram (b).

ImageJ does provide a special image type (`BinaryProcessor`) for binary images, but these are actually implemented as 8-bit intensity images (just like ordinary intensity images) using the values 0 and 255. ImageJ also provides the `ImageProcessor` method `threshold(int level)`, with `level` $\equiv a_{\mathrm{th}}$, to perform this operation, which can also be invoked through the Image→Adjust→Threshold menu (see Fig. 4.1 for an example). Thresholding affects the histogram by splitting and merging the distribution into two entries at positions a_0 and a_1, as illustrated in Fig. 4.2.

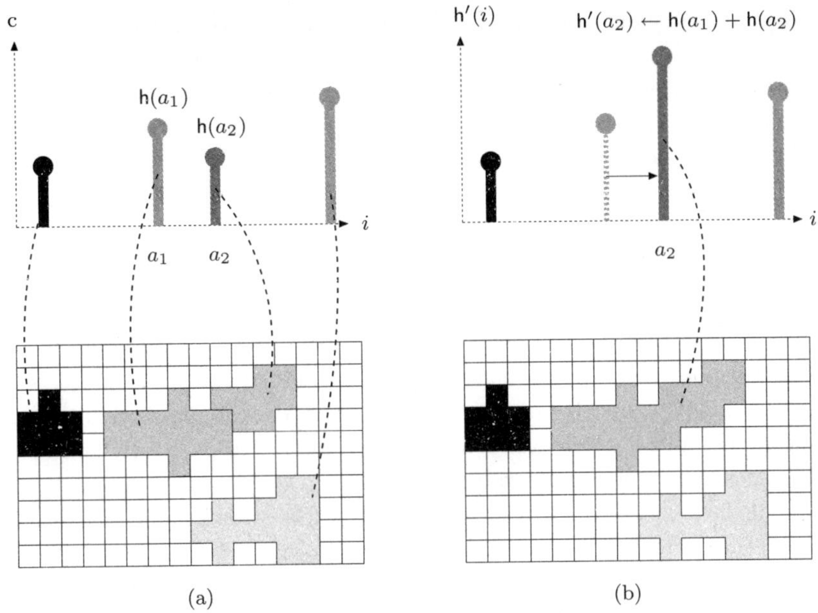

Figure 4.3 Histogram entries map to *sets* of pixels of the same value. If a histogram line is moved as a result of some point operations, then all pixels in the corresponding set are equally modified (a). If, due to this operation, two histogram lines $\mathsf{h}(a_1)$, $\mathsf{h}(a_2)$ coincide on the same index, the two corresponding pixel sets join and the contained pixels become undiscernable (b).

4.2 Point Operations and Histograms

We have already seen that the effects of a point operation on the image's histogram are quite easy to predict in some cases. For example, increasing the brightness of an image by a constant value shifts the entire histogram to the right, raising the contrast widens it, and inverting the image flips the histogram. Although this appears rather simple, it may be useful to look a bit more closely at the relationship between point operations and the resulting changes in the histogram.

As the illustration in Fig. 4.3 shows, every entry (bar) at some position i in the histogram maps to a *set* (of size $\mathsf{h}(i)$) containing all image pixels whose values are exactly i.[2] If a particular histogram line is *shifted* as a result of some point operation, then of course all pixels in the corresponding set are equally modified and vice versa. So what happens when a point operation (e. g.,

---

[2] Of course this is only true for ordinary histograms with an entry for every single intensity value. If *binning* is used (see Sec. 3.4.1), each histogram entry maps to pixels within a certain *range* of values.

reducing image contrast) causes two previously separated histogram lines to fall together at the same position i? The answer is that the corresponding pixel sets are *merged* and the new common histogram entry is the sum of the two (or more) contributing entries (i. e., the size of the combined set). At this point, the elements in the merged set are no longer distinguishable (or separable), so this operation may have (perhaps unintentionally) caused an irreversible reduction of dynamic range and thus a permanent loss of information in that image.

4.3 Automatic Contrast Adjustment

Automatic contrast adjustment ("auto-contrast") is a point operation whose task is to modify the pixels such that the available range of values is fully covered. This is done by mapping the current darkest and brightest pixels to the lowest and highest available intensity values, respectively, and linearly distributing the intermediate values.

Let us assume that a_{low} and a_{high} are the lowest and highest pixel values found in the current image, whose full intensity range is $[a_{\min}, a_{\max}]$. To stretch the image to the full intensity range (see Fig. 4.4), we first map the smallest pixel value a_{low} to zero, subsequently increase the contrast by the factor $(a_{\max} - a_{\min})/(a_{\text{high}} - a_{\text{low}})$, and finally shift to the target range by adding $a_{\min}$. The mapping function for the auto-contrast operation is thus defined as

$$f_{\text{ac}}(a) = a_{\min} + \big(a - a_{\text{low}}\big) \cdot \frac{a_{\max} - a_{\min}}{a_{\text{high}} - a_{\text{low}}}, \tag{4.6}$$

provided that $a_{\text{high}} \neq a_{\text{low}}$; i. e., the image contains at least *two* different pixel values. For an 8-bit image with $a_{\min} = 0$ and $a_{\max} = 255$, the function in Eqn. (4.6) simplifies to

$$f_{\text{ac}}(a) = (a - a_{\text{low}}) \cdot \frac{255}{a_{\text{high}} - a_{\text{low}}}. \tag{4.7}$$

The target range $[a_{\min}, a_{\max}]$ need not be the maximum available range of values but can be any interval to which the image should be mapped. Of course the method can also be used to reduce the image contrast to a smaller range. Figure 4.5 (b) shows the effects of an auto-contrast operation on the corresponding histogram, where the linear stretching of the intensity range results in regularly spaced gaps in the new distribution.

4.4 Modified Auto-Contrast

In practice, the mapping function in Eqn. (4.6) could be strongly influenced by only a few extreme (low or high) pixel values, which may not be representative of the main image content. This can be avoided to a large extent by "saturating"

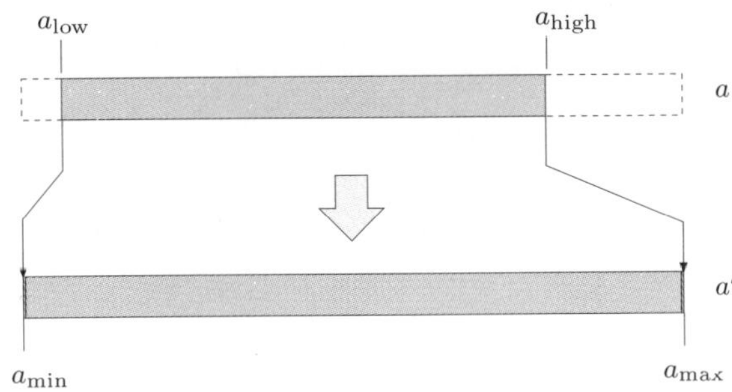

Figure 4.4 Auto-contrast operation according to Eqn. (4.6). Original pixel values a in the range $[a_{low}, a_{high}]$ are mapped linearly to the target range $[a_{min}, a_{max}]$.

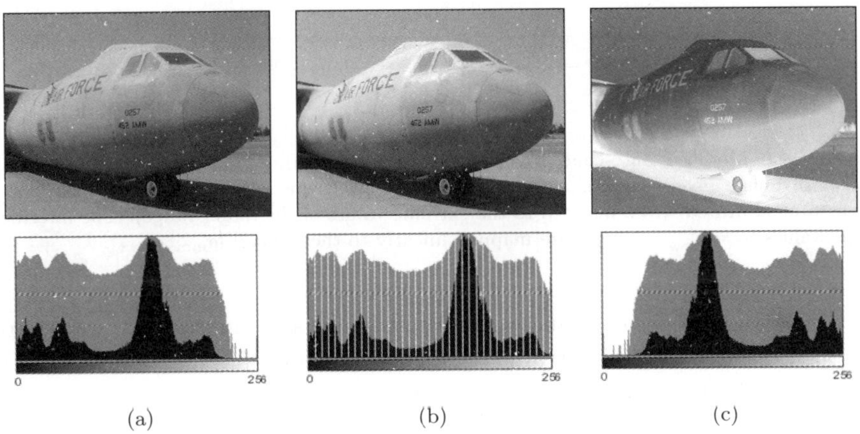

Figure 4.5 Effects of auto-contrast and inversion operations on the resulting histograms. Original image (a), result of auto-contrast operation (b), and inversion (c). The histogram entries are shown both linearly (black bars) and logarithmically (gray bars).

a fixed percentage (s_{low}, s_{high}) of pixels at the upper and lower ends of the target intensity range. To accomplish this, we determine two limiting values a'_{low}, a'_{high} such that a predefined quantile q_{low} of all pixel values in the image I are smaller than a'_{low} and another quantile q_{high} of the values are greater than a'_{high} (Fig. 4.6). The values a'_{low}, a'_{high} depend on the image content and can be easily obtained from the image's cumulative histogram[3] $\mathsf{H}(i)$:

$$a'_{low} = \min\{ i \mid \mathsf{H}(i) \geq M \cdot N \cdot q_{low} \}, \tag{4.8}$$
$$a'_{high} = \max\{ i \mid \mathsf{H}(i) \leq M \cdot N \cdot (1 - q_{high}) \}, \tag{4.9}$$

where $0 \leq q_{low}, q_{high} \leq 1$, $q_{low} + q_{high} \leq 1$, and $M \cdot N$ is the number of pixels in the image. All pixel values *outside* (and including) a'_{low} and a'_{high} are mapped

[3] See Sec. 3.6.

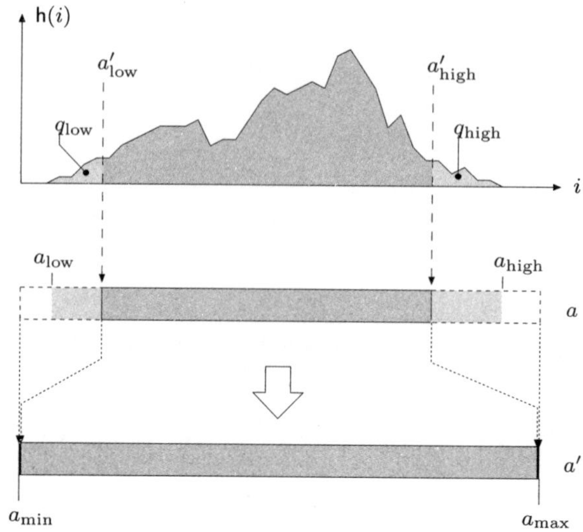

Figure 4.6 Modified auto-contrast operation (Eqn. (4.10)). Predefined quantiles (q_{low}, q_{high}) of image pixels—shown as dark areas at the left and right ends of the histogram $\mathsf{h}(i)$—are "saturated" (i. e., mapped to the extreme values of the target range). The intermediate values ($a = a'_{low} \ldots a'_{high}$) are mapped linearly to the interval $[a_{min}, a_{max}]$.

to the extreme values a_{min} and a_{max}, respectively, and intermediate values are mapped linearly to the interval $[a_{min}, a_{max}]$. The mapping function $f_{mac}()$ for the modified auto-contrast operation can thus be defined as

$$f_{mac}(a) = \begin{cases} a_{min} & \text{for } a \leq a'_{low} \\ a_{min} + \left(a - a'_{low}\right) \cdot \dfrac{a_{max} - a_{min}}{a'_{high} - a'_{low}} & \text{for } a'_{low} < a < a'_{high} \\ a_{max} & \text{for } a \geq a'_{high}. \end{cases} \tag{4.10}$$

Using this formulation, the mapping to minimum and maximum intensities does not depend on singular extreme pixels only but can be based on a representative set of pixels. Usually the same value is taken for both upper and lower quantiles (i. e., $q_{low} = q_{high} = q$), with $q = 0.005 \ldots 0.015$ ($0.5 \ldots 1.5$ %) being common values. For example, the auto-contrast operation in Adobe Photoshop saturates 0.5 % ($q = 0.005$) of all pixels at both ends of the intensity range. Auto-contrast is a frequently used point operation and thus available in practically any image-processing software. ImageJ implements the modified auto-contrast operation as part of the Brightness/Contrast and Image→Adjust menus (Auto button), shown in Fig. 4.7.

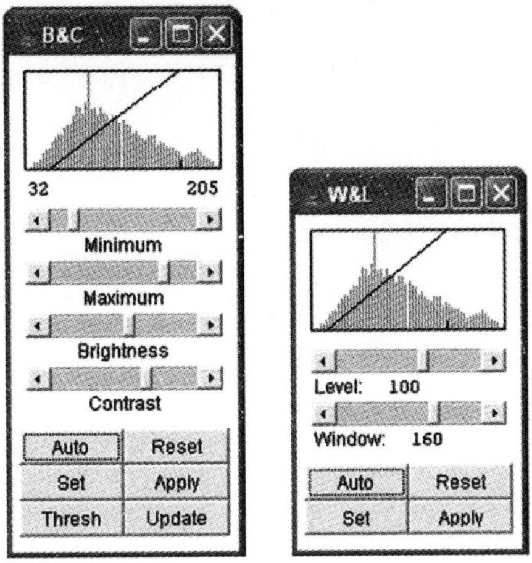

Figure 4.7 ImageJ's Brightness/Contrast tool (left) and Window/Level tool (right) can be invoked through the Image→Adjust menu. The Auto button displays the result of a modified auto-contrast operation. Apply must be hit to actually modify the image.

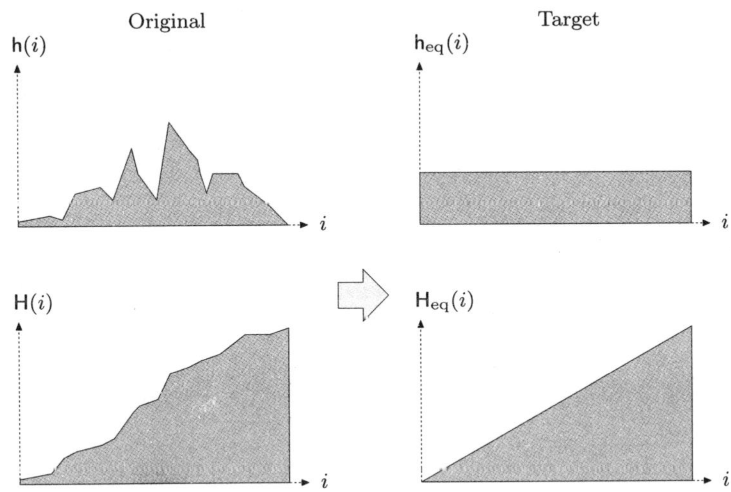

Figure 4.8 Histogram equalization. The idea is to find and apply a point operation to the image (with original histogram h) such that the histogram h_{eq} of the modified image approximates a *uniform* distribution (top). The cumulative target histogram H_{eq} must thus be approximately wedge-shaped (bottom).

4.5 Histogram Equalization

A frequent task is to adjust two different images in such a way that their resulting intensity distributions are similar, for example to use them in a print publication or to make them easier to compare. The goal of histogram equalization is to find and apply a point operation such that the histogram of the modified image approximates a *uniform* distribution (see Fig. 4.8). Since the

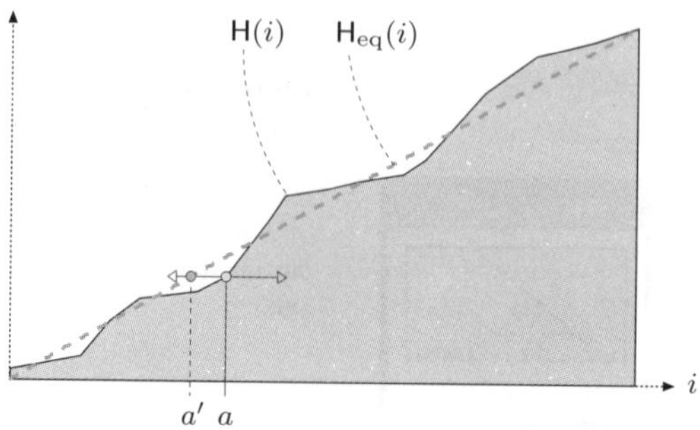

Figure 4.9 Histogram equalization on the cumulative histogram. A suitable point operation $a' \leftarrow f_{\mathrm{eq}}(a)$ shifts each histogram line from its original position a to a' (left or right) such that the resulting cumulative histogram H_{eq} is approximately linear.

histogram is a discrete distribution and homogeneous point operations can only shift and merge (but never split) histogram entries, we can only obtain an approximate solution in general. In particular, there is no way to eliminate or decrease individual peaks in a histogram, and a truly uniform distribution is thus impossible to reach. Based on point operations, we can thus modify the image only to the extent that the resulting histogram is *approximately* uniform. The question is how good this approximation can be and exactly which point operation (which clearly depends on the image content) we must apply to achieve this goal.

We may get a first idea by observing that the *cumulative* histogram (Sec. 3.6) of a uniformly distributed image is a linear ramp (wedge), as shown in Fig. 4.8. So we can reformulate the goal as finding a point operation that shifts the histogram lines such that the resulting cumulative histogram is approximately linear, as illustrated in Fig. 4.9.

The desired point operation $f_{\mathrm{eq}}()$ is simply obtained from the cumulative histogram H of the original image as[4]

$$f_{\mathrm{eq}}(a) = \left\lfloor \mathsf{H}(a) \cdot \frac{K-1}{MN} \right\rfloor , \tag{4.11}$$

for an image of size $M \times N$ with pixel values a in the range $[0, K-1]$. The resulting function $f_{\mathrm{eq}}(a)$ in Eqn. (4.11) is monotonically increasing, because $\mathsf{H}(a)$ is monotonic and K, M, N are all positive constants. In the (unusual) case where an image is already uniformly distributed, linear histogram equalization should not modify that image any further. Also, repeated applications of linear histogram equalization should not make any changes to the image after the first time. Both requirements are fulfilled by the formulation in Eqn. (4.11).

[4] For a derivation, see, e. g., [17, p. 173].

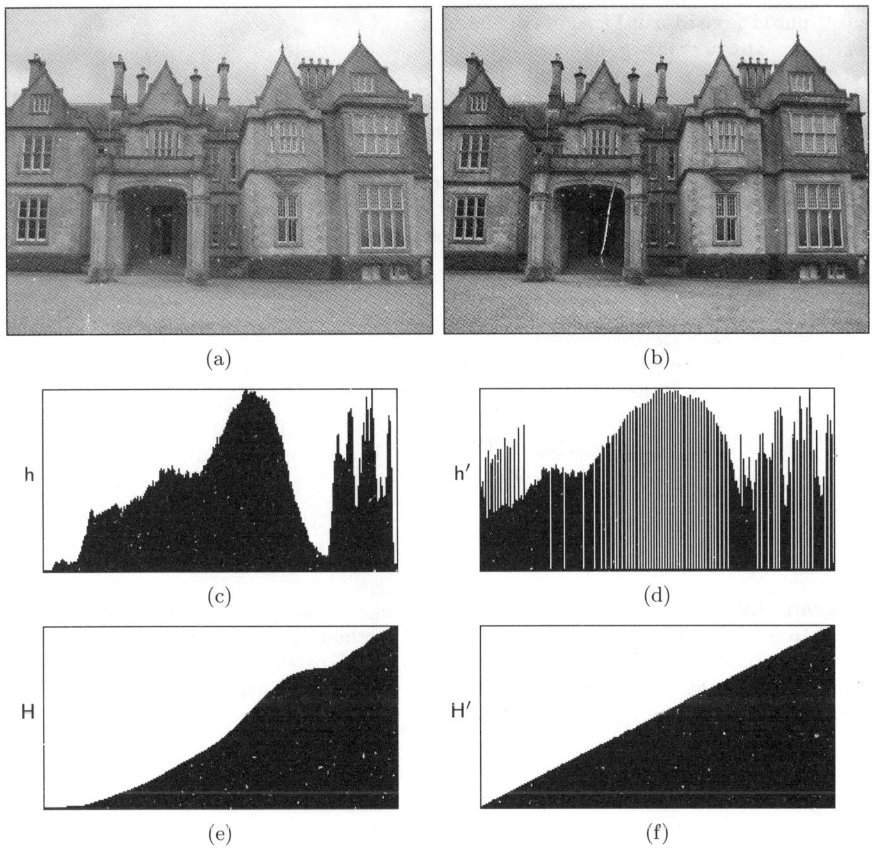

Figure 4.10 Linear histogram equalization (example). Original image I (a) and modified image I' (b), corresponding histograms h, h' (c, d), and cumulative histograms H, H' (e, f). The resulting cumulative histogram H' (f) approximates a uniformly distributed image. Notice that new peaks are created in the resulting histogram h' (d) by merging original histogram cells, particularly in the lower and upper intensity ranges.

Program 4.2 lists the Java code for a sample implementation of linear histogram equalization. An example demonstrating the effects on the image and the histograms is shown in Fig. 4.10.

Notice that for "inactive" pixel values i (i. e., pixel values that do not appear in the image, with $\mathsf{h}(i) = 0$), the corresponding entries in the cumulative histogram $\mathsf{H}(i)$ are either zero or identical to the neighboring entry $\mathsf{H}(i-1)$. Consequently a contiguous range of zero values in the histogram $\mathsf{h}(i)$ corresponds to a constant (i. e., flat) range in the cumulative histogram $\mathsf{H}(i)$, and the function $f_{\text{eq}}(a)$ maps all "inactive" intensity values within such a range to the next lower "active" value. This effect is not relevant, however, since the image contains no such pixels anyway. Nevertheless, a linear histogram equalization

```
1  public void run(ImageProcessor ip) {
2      int w = ip.getWidth();
3      int h = ip.getHeight();
4      int M = w * h;    // total number of image pixels
5      int K = 256;      // number of intensity values
6
7      // compute the cumulative histogram:
8      int[] H = ip.getHistogram();
9      for (int j = 1; j < H.length; j++) {
10         H[j] = H[j-1] + H[j];
11     }
12
13     // equalize the image:
14     for (int v = 0; v < h; v++) {
15         for (int u = 0; u < w; u++) {
16             int a = ip.get(u, v);
17             int b = H[a] * (K-1) / M;
18             ip.set(u, v, b);
19         }
20     }
21 }
```

Program 4.2 Linear histogram equalization (ImageJ plugin). First the histogram of the image `ip` is obtained using the standard ImageJ method `ip.getHistogram()` in line 8. In line 10, the cumulative histogram is computed "in place" based on the recursive definition in Eqn. (3.6). The `int` division in line 17 implicitly performs the required floor ($\lfloor\ \rfloor$) operation by truncation.

may (and typically will) cause histogram lines to merge and consequently lead to a loss of dynamic range (see also Sec. 4.2).

This or a similar form of linear histogram equalization is implemented in almost any image-processing software. In ImageJ it can be invoked interactively through the Process→Enhance Contrast menu (option Equalize). To avoid extreme contrast effects, the histogram equalization in ImageJ by default[5] cumulates the *square root* of the histogram entries using a modified cumulative histogram of the form

$$\tilde{\mathsf{H}}(i) = \sum_{j=0}^{i} \sqrt{\mathsf{h}(j)}\,. \tag{4.12}$$

4.6 Histogram Specification

Although widely implemented, the goal of linear histogram equalization—a uniform distribution of intensity values (as described in the previous section)—appears rather ad hoc, since good images virtually never show such a distri-

[5] The "classic" (linear) approach, as given by Eqn. (3.5), is used when simultaneously keeping the Alt key pressed.

bution. In most real images, the distribution of the pixel values is not even remotely uniform but is usually more similar, if at all, to perhaps a Gaussian distribution. The images produced by linear equalization thus usually appear quite unnatural, which renders the technique practically useless.

Histogram specification is a more general technique that modifies the image to match an arbitrary intensity distribution, including the histogram of a given image. This is particularly useful, for example, for adjusting a set of images taken by different cameras or under varying exposure or lighting conditions to give a similar impression in print production or when displayed. Similar to histogram equalization, this process relies on the alignment of the cumulative histograms by applying a homogeneous point operation. To be independent of the image size (i. e., the number of pixels), we first define *normalized* distributions, which we use in place of the original histograms.

4.6.1 Frequencies and Probabilities

The value in each histogram cell describes the observed frequency of the corresponding intensity value, i. e., the histogram is a discrete *frequency distribution*. For a given image I of size $M \times N$, the sum of all histogram entries $\mathsf{h}(i)$ equals the number of image pixels,

$$\sum_{i=0}^{K-1} \mathsf{h}(i) = M \cdot N. \tag{4.13}$$

The associated *normalized* histogram

$$\mathsf{p}(i) = \frac{\mathsf{h}(i)}{MN}, \tag{4.14}$$

for $0 \leq i < K$, is usually interpreted as the *probability distribution* or *probability density function* (pdf) of a random process, where $\mathsf{p}(i)$ is the probability for the occurrence of the pixel value i. The cumulative probability of i being any possible value is 1, and the distribution p must thus satisfy

$$\sum_{i=0}^{K-1} \mathsf{p}(i) = 1. \tag{4.15}$$

The statistical counterpart to the cumulative histogram H (Eqn. (3.5)) is the discrete *distribution function* $\mathsf{P}()$ (also called the *cumulative distribution function* or cdf), with

$$\begin{aligned} \mathsf{P}(i) &= \frac{\mathsf{H}(i)}{\mathsf{H}(K-1)} = \frac{\mathsf{H}(i)}{MN} = \sum_{j=0}^{i} \frac{\mathsf{h}(j)}{MN} \\ &= \sum_{j=0}^{i} \mathsf{p}(j), \qquad \text{for } 0 \leq i < K. \end{aligned} \tag{4.16}$$

Algorithm 4.1 Computation of the cumulative distribution function (cdf) $\mathsf{P}()$ from a given histogram h of length K. See Prog. 4.3 (p. 75) for the corresponding Java implementation.

```
CDF(h)
    Returns the cumulative distribution function P(i) ∈ [0,1] for a given
    histogram h(i), with i = 0,... K−1.
    Let K ← Size(h)
    Let n ← Σ_{i=0}^{K−1} h(i)
    Create table P of size K
    Let c ← 0
    for i ← 0...(K−1) do
        c ← c + h(i)                          ▷ cumulate histogram values
        P(i) ← c/n
    return P.
```

The computation of the cdf from a given histogram h is outlined in Alg. 4.1. The resulting function $\mathsf{P}(i)$ is (like the cumulative histogram) monotonically increasing and, in particular,

$$\mathsf{P}(0) = \mathsf{p}(0) \qquad \text{and} \qquad \mathsf{P}(K-1) = \sum_{i=0}^{K-1} \mathsf{p}(i) = 1. \tag{4.17}$$

This statistical formulation implicitly treats the generation of images as a random process whose exact properties are mostly unknown.[6] However, the process is usually assumed to be homogeneous (independent of the image position); i. e., each pixel value is the result of a "random experiment" on a single random variable i. The observed frequency distribution given by the histogram $\mathsf{h}(i)$ serves as a (coarse) estimate of the probability distribution $\mathsf{p}(i)$ of this random variable.

4.6.2 Principle of Histogram Specification

The goal of histogram specification is to modify a given image I_A by some point operation such that its distribution function P_A matches a *reference distribution* P_R as closely as possible. We thus look for a mapping function

$$a' = f_{\text{hs}}(a) \tag{4.18}$$

[6] Statistical modeling of the image generation process has a long tradition (see, e. g., [25, Ch. 2]).

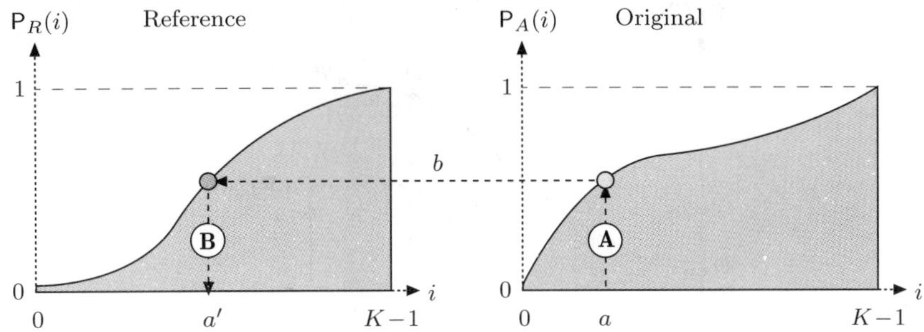

Figure 4.11 Principle of histogram specification. Given is the reference distribution P_R (left) and the distribution function for the original image P_A (right). The result is the mapping function $f_{\mathrm{hs}}: a \rightarrow a'$ for a point operation, which replaces each pixel a in the original image I_A by a modified value a'. The process has two main steps: Ⓐ For each pixel value a, determine $b = \mathsf{P}_A(a)$ from the right distribution function. Ⓑ a' is then found by inverting the left distribution function as $a' = \mathsf{P}_R^{-1}(b)$. In summary, the result is $f_{\mathrm{hs}}(a) = a' = \mathsf{P}_R^{-1}(\mathsf{P}_A(a))$.

to convert the original image I_A to a new image $I_{A'}$ by a point operation such that

$$\mathsf{P}_{A'}(i) \approx \mathsf{P}_R(i) \qquad \text{for} \ \ 0 \leq i < K. \tag{4.19}$$

As illustrated in Fig. 4.11, the desired mapping f_{hs} is found by combining the two distribution functions P_R and P_A (see [17, p. 180] for details). For a given pixel value a in the original image, we get the new pixel value a' as

$$a' = \mathsf{P}_R^{-1}\big(\mathsf{P}_A(a)\big), \tag{4.20}$$

and thus the mapping f_{hs} (Eqn. (4.18)) is obtained as

$$f_{\mathrm{hs}}(a) = a' = \mathsf{P}_R^{-1}\big(\mathsf{P}_A(a)\big) \tag{4.21}$$

for $0 \leq a < K$. This of course assumes that $\mathsf{P}_R(i)$ is invertible; i. e., that the function $\mathsf{P}_R^{-1}(b)$ exists for $b \in [0,1]$.

4.6.3 Adjusting to a Piecewise Linear Distribution

If the reference distribution P_R is given as a continuous, invertible function, then the mapping function f_{hs} can be obtained from Eqn. (4.21) without any difficulty. In practice, it is convenient to specify the (synthetic) reference distribution as a *piecewise linear* function $\mathsf{P}_L(i)$; i. e., as a sequence of $N+1$ coordinate pairs

$$\mathcal{L} = [\langle a_0, q_0 \rangle, \langle a_1, q_1 \rangle, \ldots \langle a_k, q_k \rangle, \ldots \langle a_N, q_N \rangle],$$

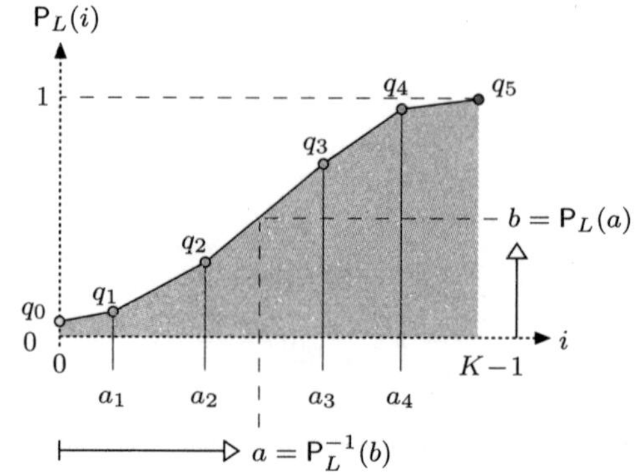

Figure 4.12 Piecewise linear reference distribution. The function $\mathsf{P}_L(i)$ is specified by $N = 5$ control points $\langle 0, q_0\rangle$, $\langle a_1, q_1\rangle$, ... $\langle a_4, q_4\rangle$, with $a_k < a_{k+1}$ and $q_k < q_{k+1}$. The final point q_5 is fixed at $\langle K-1, 1\rangle$.

each consisting of an intensity value a_k and the corresponding function value q_k (with $0 \leq a_k < K$, $a_k < a_{k+1}$, and $0 \leq q_k < 1$). The two endpoints $\langle a_0, q_0\rangle$ and $\langle a_N, q_N\rangle$ are fixed at

$$\langle 0, q_0\rangle \qquad \text{and} \qquad \langle K-1, 1\rangle,$$

respectively. To be invertible, the function must also be strictly monotonic; i. e., $q_k < q_{k+1}$ for $0 \leq k < N$. Figure 4.12 shows an example for such a function, which is specified by $N = 5$ variable points $(q_0, \ldots q_4)$ and a fixed end point q_5 and thus consists of $N = 5$ linear segments. The reference distribution can of course be specified at an arbitrary accuracy by inserting additional control points.

The intermediate values of $\mathsf{P}_L(i)$ are obtained by linear interpolation between the control points as

$$\mathsf{P}_L(i) = \begin{cases} q_m + (i - a_m)\cdot\dfrac{(q_{m+1} - q_m)}{(a_{m+1} - a_m)} & \text{for } 0 \leq i < K-1 \\ 1 & \text{for } i = K-1, \end{cases} \tag{4.22}$$

where $m = \max\{j \in [0, N-1] \mid a_j \leq i\}$ is the index of the line segment $\langle a_m, q_m\rangle \to \langle a_{m+1}, q_{m+1}\rangle$, which overlaps the position i. For instance, in the example in Fig. 4.12, the point a lies within the segment that starts at point $\langle a_2, q_2\rangle$; i. e., $m = 2$.

For the histogram specification according to Eqn. (4.21), we also need the *inverse* distribution function $\mathsf{P}_L^{-1}(b)$ for $b \in [0, 1]$. As we see from the example in Fig. 4.12, the function $\mathsf{P}_L(i)$ is in general not invertible for values $b < \mathsf{P}_L(0)$. We can fix this problem by mapping all values $b < \mathsf{P}_L(0)$ to zero and thus obtain a "semi-inverse" of the reference distribution in Eqn. (4.22) as

$$\mathsf{P}_L^{-1}(b) = \begin{cases} 0 & \text{for } 0 \leq b < \mathsf{P}_L(0) \\ a_n + (b - q_n) \cdot \dfrac{(a_{n+1} - a_n)}{(q_{n+1} - q_n)} & \text{for } \mathsf{P}_L(0) \leq b < 1 \\ K-1 & \text{for } b \geq 1. \end{cases} \tag{4.23}$$

Here $n = \max\{j \in \{0, \ldots N-1\} \mid q_j \leq b\}$ is the index of the line segment $\langle a_n, q_n \rangle \rightarrow \langle a_{n+1}, q_{n+1} \rangle$, which overlaps the argument value b. The required mapping function f_{hs} for adapting a given image with intensity distribution P_A is finally specified, analogous to Eqn. (4.21), as

$$f_{\text{hs}}(a) = \mathsf{P}_L^{-1}\big(\mathsf{P}_A(a)\big) \qquad \text{for } 0 \leq a < K. \tag{4.24}$$

The whole process of computing the pixel mapping function for a given image (histogram) and a piecewise linear target distribution is summarized in Alg. 4.2. A real example is shown in Fig. 4.14 (Sec. 4.6.5).

4.6.4 Adjusting to a Given Histogram (Histogram Matching)

If we want to adjust one image to the histogram of another image, the reference distribution function $\mathsf{P}_R(i)$ is not continuous and thus, in general, cannot be inverted (as required by Eqn. (4.21)). For example, if the reference distribution contains zero entries (i. e., pixel values k with probability $\mathsf{p}(k) = 0$), the corresponding cumulative distribution function P (just like the cumulative histogram) has intervals of constant value on which no inverse function value can be determined.

In the following, we describe a simple method for histogram matching that works with discrete reference distributions. The principal idea is graphically illustrated in Fig. 4.13. The mapping function f_{hs} is not obtained by inverting but by "filling in" the reference distribution function $\mathsf{P}_R(i)$. For each possible pixel value a, starting with $a = 0$, the corresponding probability $\mathsf{p}_A(a)$ is stacked layer by layer "under" the reference distribution P_R. The thickness of each horizontal bar for a equals the corresponding probability $\mathsf{p}_A(a)$. The bar for a particular intensity value a with thickness $\mathsf{p}_A(a)$ runs from right to left, down to position a', where it hits the reference distribution P_R. This position a' corresponds to the new pixel value to which a should be mapped.

Since the sum of all probabilities p_A and the maximum of the distribution function P_R are both 1 (i. e., $\sum_i \mathsf{p}_A(i) = \max_i \mathsf{P}_R(i) = 1$), all horizontal bars will exactly fit underneath the function P_R. One may also notice in Fig. 4.13 that the distribution value resulting at a' is identical to the cumulated probability $\mathsf{P}_A(a)$. Given some intensity value a, it is therefore sufficient to find the

Algorithm 4.2 Histogram specification using a piecewise linear reference distribution. Given is the histogram h_A of the original image and a piecewise linear reference distribution function, specified as a sequence of N control points $\mathcal{L}_R$. The discrete mapping function f_{hs} for the corresponding point operation is returned.

MatchPiecewiseLinearHistogram(h_A, $\mathcal{L}_R$)

h_A: histogram of the original image I_A.
$\mathcal{L}_R$: reference distribution function, given as a sequence of $N+1$ control points $\mathcal{L}_R = [\langle a_0, q_0\rangle, \langle a_1, q_1\rangle, \ldots \langle a_N, q_N\rangle]$, with $0 \leq a_k < K$ and $0 \leq q_k \leq 1$.

Returns a discrete pixel mapping function $f_{\mathrm{hs}}(a)$ for modifying the original image I_A.

- Let $K \leftarrow \mathrm{Size}(\mathsf{h}_A)$
- Let $\mathsf{P}_A \leftarrow \mathrm{Cdf}(\mathsf{h}_A)$ ▷ cdf for h_A (Alg. 4.1)
- Create a table $f_{\mathrm{hs}}[\,]$ of size K ▷ mapping function f_{hs}
- **for** $a \leftarrow 0 \ldots (K-1)$ **do**
  - $b \leftarrow \mathsf{P}_A(a)$
  - **if** $(b \leq q_0)$ **then**
    - $a' \leftarrow 0$
  - **else if** $(b \geq 1)$ **then**
    - $a' \leftarrow K-1$
  - **else**
    - $n \leftarrow N-1$
    - **while** $(n \geq 0) \wedge (q_n > b)$ **do** ▷ find line segment in $\mathcal{L}_R$
      - $n \leftarrow n-1$
    - $a' \leftarrow a_n + (b-q_n) \cdot \dfrac{(a_{n+1} - a_n)}{(q_{n+1} - q_n)}$ ▷ see Eqn. (4.23)
  - $f_{\mathrm{hs}}[a] \leftarrow a'$
- **return** f_{hs}.

minimum value a', where the reference distribution $\mathsf{P}_R(a')$ is greater than or equal to the cumulative probability $\mathsf{P}_A(a)$; i.e.,

$$f_{\mathrm{hs}}(a) = a' = \min\{\, j \mid (0 \leq j < K) \wedge \big(\mathsf{P}_A(a) \leq \mathsf{P}_R(j)\big)\big\}. \qquad (4.25)$$

This results in a very simple method, which is summarized in Alg. 4.3. Due to the use of normalized distribution functions, the *size* of the images involved is not relevant. The corresponding Java implementation in Prog. 4.3, consists of the method `matchHistograms()`, which accepts the original histogram (`Ha`) and the reference histogram (`Hr`) and returns the resulting mapping function (`map`) specifying the required point operation. The following code fragment

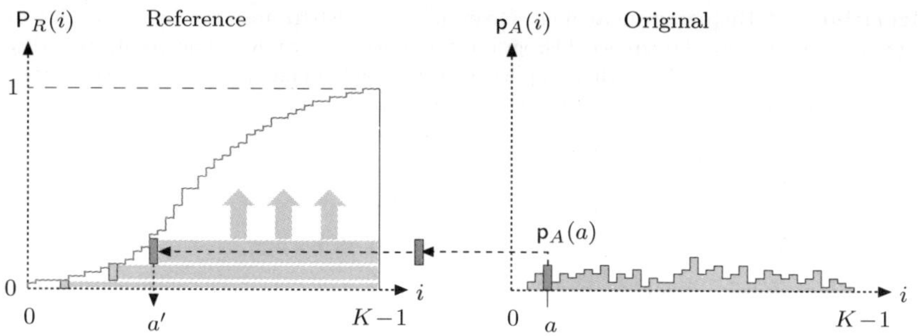

Figure 4.13 Discrete histogram specification. The reference distribution P_R (left) is "filled" layer by layer from bottom to top and from right to left. For every possible intensity value a (starting from $a = 0$), the associated probability $p_A(a)$ is added as a horizontal bar to a stack accumulated 'under" the reference distribution P_R. The bar with thickness $p_A(a)$ is drawn from right to left down to the position a', where the reference distribution P_R is reached. This value a' is the one which a should be mapped to by the function $f_{hs}(a)$.

demonstrates the use of the method `matchHistograms()` from Prog. 4.3 in an ImageJ program:

```
ImageProcessor ipA = ...      // target image I_A (to be modified)
ImageProcessor ipR = ...      // reference image I_R

int[] hA = ipA.getHistogram(); // get the histogram for I_A
int[] hR = ipR.getHistogram(); // get the histogram for I_R

int[] F = matchHistograms(hA, hR); // mapping function f_hs(a)
ipA.applyTable(F);             // apply f_hs() to the target image I_A
```

The original image `ipA` is modified in the last line by applying the mapping function f_{hs} (`F`) with the method `applyTable()` (see also p. 87).

4.6.5 Examples

Adjusting to a piecewise linear reference distribution

The first example in Fig. 4.14 shows the results of histogram specification for a continuous, piecewise linear reference distribution, as described in Sec. 4.6.3. Analogous to Fig. 4.12, the actual distribution function P_R (Fig. 4.14 (f)) is specified as a polygonal line consisting of five control points $\langle a_k, q_k \rangle$ with coordinates

$k =$	0	1	2	3	4	5
$a_k =$	0	28	75	150	210	255
$q_k =$	0.002	0.050	0.250	0.750	0.950	1.000

The resulting reference histogram (Fig. 4.14 (c)) is a step function with ranges of constant values corresponding to the linear segments of the probability density

Algorithm 4.3 Histogram matching. Given are two histograms: the histogram h_A of the target image I_A and a reference histogram h_R, both of size K. The result is a discrete mapping function $f_{\mathrm{hs}}()$ that, when applied to the target image, produces a new image with a distribution function similar to the reference histogram.

MatchHistograms(h_A, h_R)

  h_A: histogram of the target image I_A.
  h_R: reference histogram (of same size as h_A).

  Returns a discrete pixel mapping function $f_{\mathrm{hs}}(a)$ for modifying the original image I_A.

  Let $K \leftarrow \mathrm{Size}(\mathsf{h}_A)$
  Let $\mathsf{P}_A \leftarrow \mathrm{CDF}(\mathsf{h}_A)$ ▷ cdf for h_A (Alg. 4.1)
  Let $\mathsf{P}_R \leftarrow \mathrm{CDF}(\mathsf{h}_R)$ ▷ cdf for h_R (Alg. 4.1)
  Create a table $f_{\mathrm{hs}}[\,]$ of size K ▷ pixel mapping function f_{hs}
  for $a \leftarrow 0 \ldots (K-1)$ **do**
    $j \leftarrow K-1$
    repeat
      $f_{\mathrm{hs}}[a] \leftarrow j$
      $j \leftarrow j-1$
    while $(j \geq 0) \wedge (\mathsf{P}_A(a) \leq \mathsf{P}_R(j))$
  return f_{hs}.

function. As expected, the *cumulative* probability function for the modified image (Fig. 4.14 (h)) is quite close to the reference function in Fig. 4.14 (f), while the resulting *histogram* (Fig. 4.14 (e)) shows little similarity with the reference histogram (Fig. 4.14 (c)). However, as discussed earlier, this is all we can expect from a homogeneous point operation.

Adjusting to an arbitrary reference histogram

In this case, the reference distribution is not given as a continuous function but specified by a discrete histogram. We thus use the method described in Sec. 4.6.4 to compute the required mapping functions. The examples in Fig. 4.15 demonstrate this technique using synthetic reference histograms whose shape is approximately Gaussian.

The target image (Fig. 4.15 (a)) used here was chosen intentionally for its poor quality, manifested by an extremely unbalanced histogram (Fig. 4.15 (f)). The histograms of the modified images thus naturally show little resemblance to a Gaussian. However, the resulting *cumulative* histograms (Fig. 4.15 (j, k)) match nicely with the integral of the corresponding Gaussians (Fig. 4.15 (d, e)),

```
int[] matchHistograms (int[] hA, int[] hR) {
  // hA ... histogram h_A of target image I_A
  // hR ... reference histogram h_R
  // returns the mapping function f_hs() to be applied to image I_A

  int K = hA.length;           // hA, hR must be of length K

  double[] PA = Cdf(hA);       // get CDF of histogram h_A
  double[] PR = Cdf(hR);       // get CDF of histogram h_R

  int[] F = new int[K];        // pixel mapping function f_hs()

  // compute mapping function f_hs()
  for (int a = 0; a < K; a++) {
    int j = K-1;
    do {
      F[a] = j;
      j--;
    } while (j>=0 && PA[a]<=PR[j]);
  }

  return F;
}
```

```
double[] Cdf (int[] h) {
  // returns the cumulative distribution function for histogram h
  int K = h.length;
  int n = 0;                         // sum all histogram values
  for (int i=0; i<K; i++) {
    n += h[i];
  }

  double[] P = new double[K];  // create cdf table P
  int c = 0;                         // cumulate histogram values
  for (int i=0; i<K; i++) {
    c += h[i];
    P[i] = (double) c / n;
  }

  return P;
}
```

Program 4.3 Histogram matching (Java implementation of Alg. 4.3). The method `matchHistograms()` computes the mapping function `F` from the target histogram `Ha` and the reference histogram `Hr` (see Eqn. (4.25)). The method `Cdf()` computes the cumulative distribution function (cdf) for a given histogram (Eqn. (4.16)).

apart from the unavoidable irregularity at the center caused by the dominant peak in the original histogram.

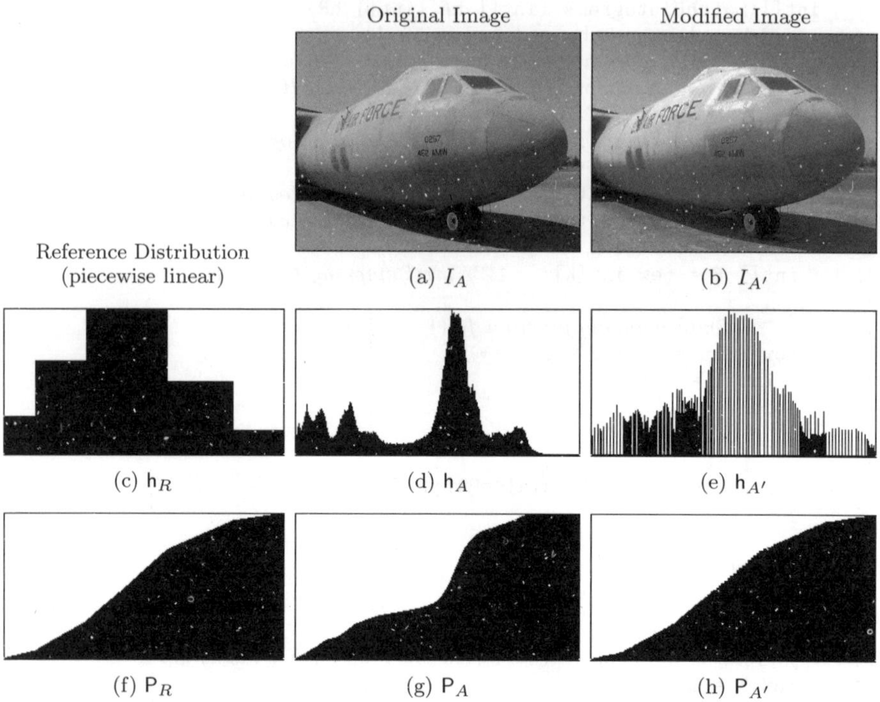

Figure 4.14 Histogram specification with a piecewise linear reference distribution. The target image I_A (a), its histogram (d), and distribution function P_A (g); the reference histogram h_R (c) and the corresponding distribution P_R (f); the modified image $I_{A'}$ (b), its histogram $\mathsf{h}_{A'}$ (e), and the resulting distribution $\mathsf{P}_{A'}$ (h).

Adjusting to another image

The third example in Fig. 4.16 demonstrates the adjustment of two images by matching their intensity histograms. One of the images is selected as the reference image I_R (Fig. 4.16 (b)) and supplies the reference histogram h_R (Fig. 4.16 (e)). The second (target) image I_A (Fig. 4.16 (a)) is modified such that the resulting cumulative histogram matches the cumulative histogram of the reference image I_R. It can be expected that the final image $I_{A'}$ (Fig. 4.16 (c)) and the reference image give a similar visual impression with regard to tonal range and distribution (assuming that both images show similar content).

Of course this method may be used to adjust multiple images to the same reference image (e. g., to prepare a series of similar photographs for a print project). For this purpose, one could either select a single representative image as a common reference or, alternatively, compute an "average" reference histogram from a set of typical images (see also Exercise 4.7).

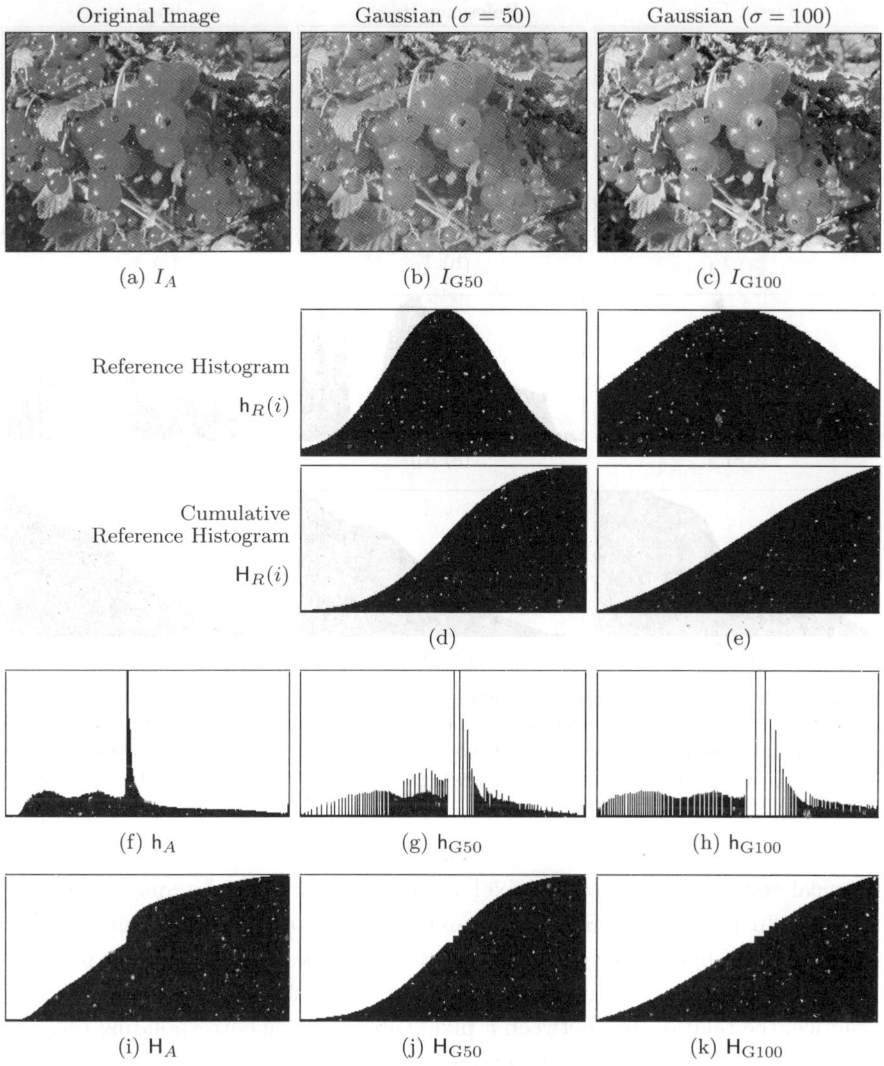

Figure 4.15 Histogram matching: adjusting to a synthetic histogram. Original image I_A (a), corresponding histogram (f), and cumulative histogram (i). Gaussian-shaped reference histograms with center $\mu = 128$ and $\sigma = 50$ (d) and $\sigma = 100$ (e), respectively. Resulting images after histogram matching, I_{G50} (b) and I_{G100} (c) with the corresponding histograms (g, h) and cumulative histograms (j, k).

4.7 Gamma Correction

We have been using the terms "intensity" and "brightness" many times without really bothering with how the numeric pixel values in our images relate to these

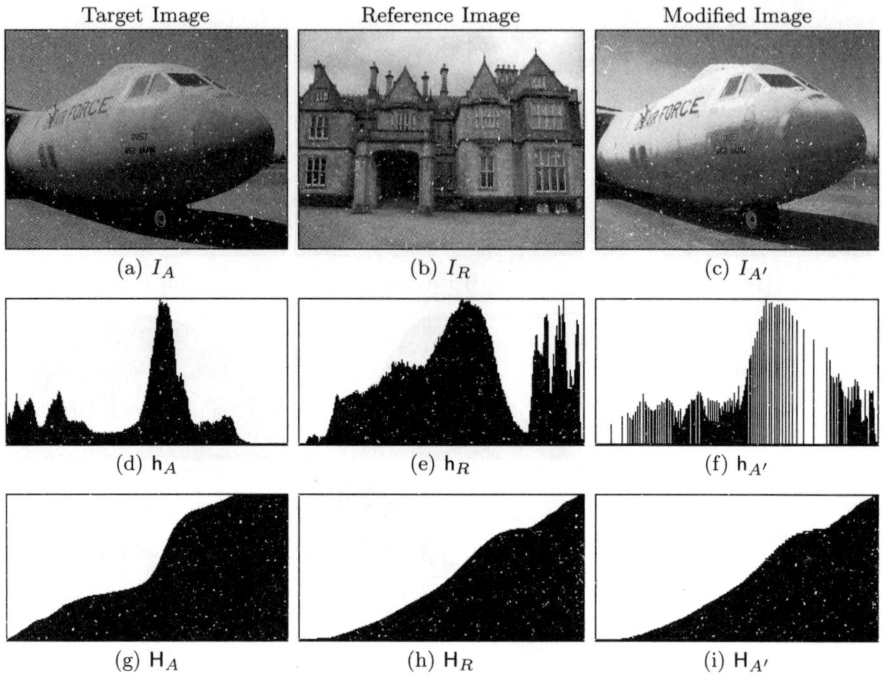

Figure 4.16 Histogram matching: adjusting to a reference image. The target image I_A (a) is modified by matching its histogram to the reference image I_R (b), resulting in the new image $I_{A'}$ (c). The corresponding histograms h_A, h_R, $\mathsf{h}_{A'}$ (d–f) and cumulative histograms H_A, H_R, $\mathsf{P}_{A'}$ (g–i) are shown. Notice the good agreement between the cumulative histograms of the reference and adjusted images (h, i).

physical concepts, if at all. A pixel value may represent the amount of light falling onto a sensor element in a camera, the photographic density of film, the amount of light to be emitted by a monitor, the number of toner particles to be deposited by a printer, or any other relevant physical magnitude. In practice, the relationship between a pixel value and the corresponding physical quantity is usually complex and almost always nonlinear. In many imaging applications, it is important to know this relationship, at least approximately, to achieve consistent and reproducible results.

When applied to digital intensity images, the ideal is to have some kind of "calibrated intensity space" that optimally matches the human perception of intensity and requires a minimum number of bits to represent the required intensity range. Gamma correction denotes a simple point operation to compensate for the transfer characteristics of different input and output devices and to map them to a unified intensity space.

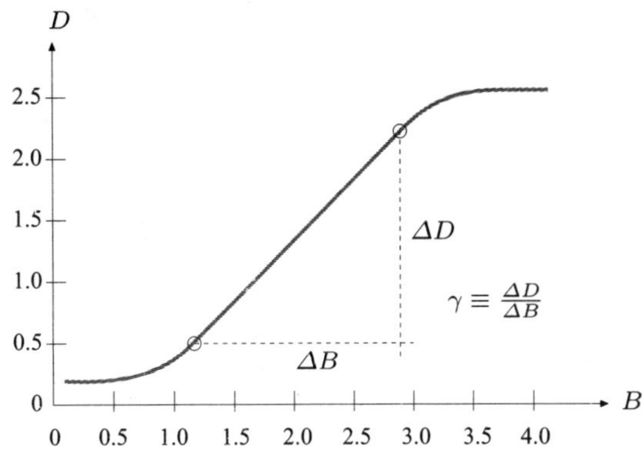

Figure 4.17 Exposure function of photographic film. With respect to the *logarithmic* light intensity B, the resulting film density D is approximately *linear* over a wide intensity range. The slope ($\Delta D/\Delta B$) of this linear section of the function specifies the "gamma" (γ) value for a particular type of photographic material.

4.7.1 Why Gamma?

The term "gamma" originates from analog photography, where the relationship between the light energy and the resulting film density is approximately logarithmic. The "exposure function" (Fig. 4.17), specifying the relationship between the *logarithmic* light intensity and the resulting film density, is therefore approximately *linear* over a wide range of light intensities. The slope of this function within this linear range is traditionally referred to as the "gamma" of the photographic material. The same term was adopted later in television broadcasting to describe the nonlinearities of the cathode ray tubes used in TV receivers, i. e., to model the relationship between the amplitude (voltage) of the video signal and the emitted light intensity. To compensate for the nonlinearities of the receivers, a "gamma correction" was (and is) applied to the TV signal once before broadcasting in order to avoid the need for costly correction measures on the receiver side.

4.7.2 Power Function

Gamma correction is based on the power function

$$f_\gamma(a) = a^\gamma \quad \text{for } a \in \mathbb{R} \text{ and } \gamma > 0, \tag{4.26}$$

where the parameter γ is called the gamma *value*. If a is constrained to the interval $[0,1]$, then—independent of γ—the value of $f_\gamma(a)$ also stays within $[0,1]$, and the function always runs through the points $(0,0)$ and $(1,1)$. In particular, $f_\gamma(a)$ is the identity function for $\gamma = 1$, as shown in Fig. 4.18. The function runs *above* the diagonal for gamma values $\gamma < 1$, and *below* it for

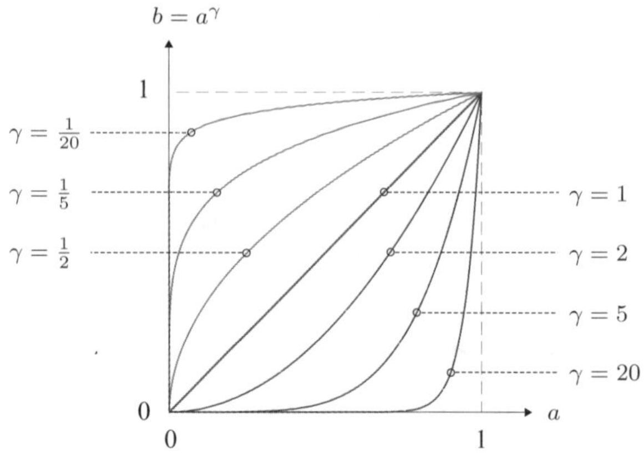

Figure 4.18 Power function $b = f_\gamma(a) = a^\gamma$ for $a \in [0, 1]$ for different gamma values.

$\gamma > 1$. Controlled by a single continuous parameter (γ), the power function can thus "imitate" both logarithmic and exponential types of functions. Within the interval $[0, 1]$, the function is continuous and strictly monotonic, and also very simple to invert as

$$a = f_\gamma^{-1}(b) = b^{1/\gamma}, \tag{4.27}$$

since $b^{1/\gamma} = (a^\gamma)^{1/\gamma} = a^1 = a$. The inverse of the power function $f_\gamma^{-1}(b)$ is thus again a power function,

$$f_\gamma^{-1}(b) = f_{\bar{\gamma}}(b) = f_{1/\gamma}(b), \tag{4.28}$$

with $\bar{\gamma} = 1/\gamma$. Thus the inverse of the power function with parameter γ is another power function with parameter $\bar{\gamma} = 1/\gamma$.

4.7.3 Real Gamma Values

The actual gamma values of individual devices are usually specified by the manufacturers based on real measurements. For example, common gamma values for CRT monitors are in the range 1.8 to 2.8, with 2.4 as a typical value. Most LCD monitors are internally adjusted to similar values. Digital video and still cameras also emulate the transfer characteristics of analog film and photographic cameras by making internal corrections to give the resulting images an accustomed "look".

In TV receivers, gamma values are standardized with 2.2 for analog NTSC and 2.8 for the PAL system (these values are theoretical; results of actual measurements are around 2.35). A gamma value of $1/2.2 \approx 0.45$ is the norm for cameras in NTSC as well as the EBU[7] standards. The current international

[7] European Broadcast Union (EBU).

standard ITU-R BT.709[8] calls for uniform gamma values of 2.5 in receivers and $1/1.956 \approx 0.51$ for cameras [15,20]. The ITU 709 standard is based on a slightly modified version of the gamma correction (see Sec. 4.7.6).

Computers usually allow adjustment of the gamma value applied to the video output signals to adapt to a wide range of different monitors. Note however that the power function $f_\gamma()$ is only a coarse approximation to the actual transfer characteristics of any device, which may also not be the same for different color channels. Thus significant deviations may occur in practice, despite the careful choice of gamma settings. Critical applications, such as prepress or high-end photography, usually require additional calibration efforts based on exactly measured device profiles (see Vol. 2 [6, Sec. 6.6.5]).

4.7.4 Applications of Gamma Correction

Let us first look at the simple example illustrated in Fig. 4.19. Assume that we use a digital camera with a nominal gamma value γ_c, meaning that its output signal s relates to the incident light intensity L as

$$S = L^{\gamma_c}. \tag{4.29}$$

To compensate the transfer characteristic of this camera (i. e., to obtain a measurement S' that is proportional to the original light intensity L), the camera signal S is subject to a gamma correction with the inverse of the camera's gamma value $\bar{\gamma}_c = 1/\gamma_c$, so

$$S' = f_{\bar{\gamma}_c}(S) = S^{1/\gamma_c}. \tag{4.30}$$

The resulting signal $S' = S^{1/\gamma_c} = (L^{\gamma_c})^{1/\gamma_c} = L^{(\gamma_c \cdot \frac{1}{\gamma_c})} = L^1$ is obviously proportional (in theory even identical) to the original light intensity L.

Although the above example is overly simplistic, it still demonstrates the general rule, which holds for output devices as well:

> The transfer characteristic of an input or output device with specified gamma value γ is compensated for by a gamma correction with $\bar{\gamma} = 1/\gamma$.

In the above, we have implicitly assumed that all values are strictly in the range $[0, 1]$, which usually is not the case in practice. When working with digital images, we have to deal with discrete pixel values; e. g., in the range $[0, 255]$ for 8-bit images. In general, performing a gamma correction

$$b = f_{\mathrm{gc}}(a, \gamma),$$

on a pixel value $a \in [0, a_{\max}]$ and a gamma value $\gamma > 0$ requires the following three steps:

[8] International Telecommunications Union (ITU).

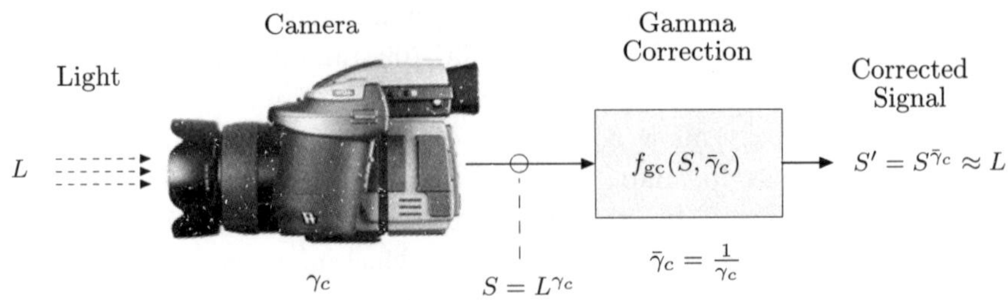

Figure 4.19 Principle of gamma correction. To compensate the output signal S produced by a camera with nominal gamma value γ_c, a gamma correction is applied with $\bar{\gamma}_c = 1/\gamma_c$. The corrected signal S' is proportional to the received light intensity L.

1. Scale a linearly to $a' \in [0, 1]$.
2. Apply the gamma correction function to a': $\quad b' \leftarrow f_\gamma(a') = a'^\gamma$.
3. Scale $b' \in [0, 1]$ linearly back to $b \in [0, a_{max}]$.

Formulated in a more compact way, the corrected pixel value b is obtained from the original value a as

$$b = f_{gc}(a, \gamma) = \left(\frac{a}{a_{max}}\right)^\gamma \cdot a_{max}. \tag{4.31}$$

Figure 4.20 illustrates the typical role of gamma correction in the digital work flow with two input (camera, scanner) and two output devices (monitor, printer), each with its individual gamma value. The central idea is to correct all images to be processed and stored in a device-independent, standardized intensity space.

4.7.5 Implementation

Program 4.4 shows the implementation of gamma correction as an ImageJ plugin for 8-bit grayscale images. The mapping function $f_{gc}(a, \gamma)$ is computed as a lookup table (`Fgc`), which is then applied to the image using the method `applyTable()` to perform the actual point operation (see also Sec. 4.8.1).

4.7.6 Modified Gamma Correction

A subtle problem with the simple power function $f_\gamma(a) = a^\gamma$ (Eqn. (4.26)) appears if we take a closer look at the *slope* of this function, expressed by its first derivative,

$$f'_\gamma(a) = \gamma \cdot a^{(\gamma-1)},$$

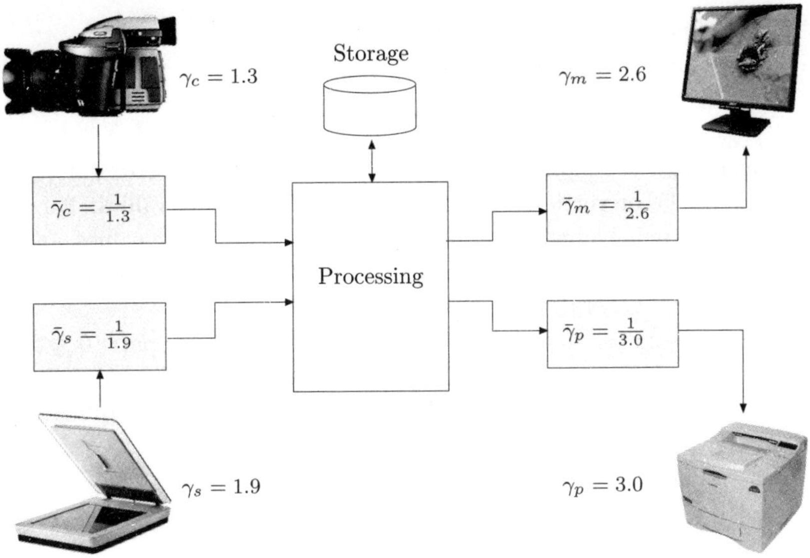

Figure 4.20 Gamma correction in the digital imaging work flow. Images are processed and stored in a "linear" intensity space, where gamma correction is used to compensate for the transfer characteristic of each input and output device. (The gamma values shown are examples only.)

```
1  public void run(ImageProcessor ip) {
2      // works for 8-bit images only
3      int K = 256;
4      int aMax = K - 1;
5      double GAMMA = 2.8;
6
7      // create a lookup table for the mapping function
8      int[] Fgc = new int[K];
9
10     for (int a = 0; a < K; a++) {
11         double aa = (double) a / aMax;   // scale to [0,1]
12         double bb = Math.pow(aa,GAMMA); // power function
13         // scale back to [0,255]:
14         int b = (int) Math.round(bb * aMax);
15         Fgc[a] = b;
16     }
17
18     ip.applyTable(Fgc);  // modify the image ip
19 }
```

Program 4.4 Gamma correction (ImageJ plugin). The corrected intensity values `b` are only computed once and stored in the lookup table `Fgc` (line 15). The gamma value `GAMMA` is constant. The actual point operation is performed by calling the ImageJ method `applyTable(Fgc)` on the image object `ip` (line 18).

which for $a = 0$ has the values

$$f'_\gamma(0) = \begin{cases} 0 & \text{for } \gamma > 1 \\ 1 & \text{for } \gamma = 1 \\ \infty & \text{for } \gamma < 1. \end{cases} \tag{4.32}$$

The tangent to the function at the origin is thus either horizontal ($\gamma > 1$), diagonal ($\gamma = 1$), or vertical ($\gamma < 1$), with no intermediate values. For $\gamma < 1$, this causes extremely high amplification of small intensity values and thus increased noise in dark image regions. Theoretically, this also means that the power function is generally not invertible at the origin.

A common solution to this problem is to replace the lower part ($0 \leq a \leq a_0$) of the power function by a linear segment with constant slope and to continue with the ordinary power function for $a > a_0$. The resulting modified gamma correction $\bar{f}_{\gamma,a_0}(a)$ is defined as

$$\bar{f}_{\gamma,a_0}(a) = \begin{cases} s \cdot a & \text{for } 0 \leq a \leq a_0 \\ (1+d) \cdot a^\gamma - d & \text{for } a_0 < a \leq 1 \end{cases} \tag{4.33}$$

$$\text{with} \quad s = \frac{\gamma}{a_0(\gamma-1) + a_0^{(1-\gamma)}} \quad \text{and} \quad d = \frac{1}{a_0^\gamma(\gamma-1)+1} - 1. \tag{4.34}$$

The function thus consists of a *linear* section (for $0 \leq a \leq a_0$) and a *nonlinear* section (for $a_0 < a \leq 1$) that connect smoothly at the transition point $a = a_0$. The linear slope s and the parameter d are determined by the requirement that the two function segments must have identical values as well as identical slopes (first derivatives) at $a = a_0$ to give a (C1) continuous function. The function in Eqn. (4.33) is thus fully specified by the two parameters a_0 and γ.

Figure 4.21 shows two examples of the modified gamma correction $\bar{f}_{\gamma,a_0}()$ with values $\gamma = 0.5$ and $\gamma = 2.0$, respectively. In both cases, the transition point is at $a_0 = 0.2$. For comparison, the figure also shows the ordinary gamma correction $f_\gamma(a)$ for the same gamma values (dashed lines), whose slope at the origin is ∞ (Fig. 4.21 (a)) and zero (Fig. 4.21 (b)), respectively.

Gamma correction in common standards

The modified gamma correction is part of several modern imaging standards. In practice, however, the values of a_0 are considerably smaller than the ones used for the illustrative examples in Fig. 4.21, and γ is chosen to obtain a good overall match to the desired correction function. For example, the ITU-BT.709 specification [20] mentioned in Sec. 4.7.3 specifies the parameters

$$\gamma = \frac{1}{2.222} \approx 0.45 \quad \text{and} \quad a_0 = 0.018,$$

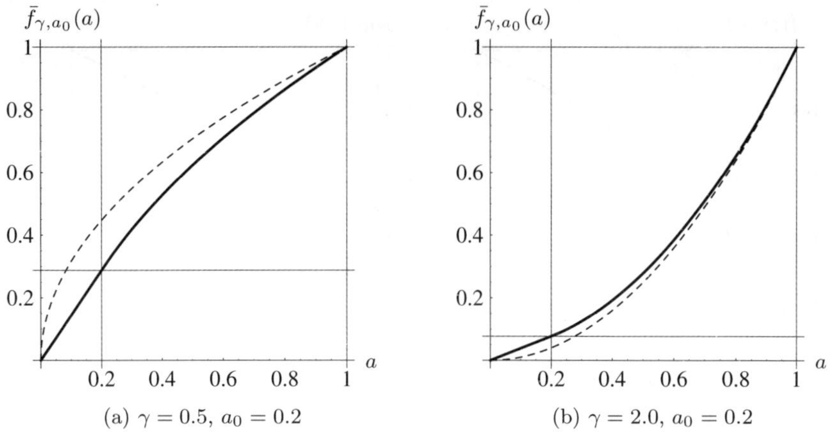

Figure 4.21 The modified gamma correction $\bar{f}_{\gamma,a_0}(a)$ consists of a linear segment with fixed slope s between $a = 0$ and $a = a_0$, followed by a power function with parameter γ (Eqn. (4.33)). The dashed lines show the ordinary power functions for the same gamma values.

Table 4.1 Gamma correction parameters for the ITU and sRGB standards Eqns. (4.33) and (4.34).

Standard	Nominal Gamma Value γ	a_0	s	d	Effective Gamma Value γ_{eff}
ITU BT.709	$1/2.222 \approx 0.450$	0.01800	4.5068	0.09915	$1/1.956 \approx 0.511$
sRGB	$1/2.400 \approx 0.417$	0.00304	12.9231	0.05500	$1/2.200 \approx 0.455$

with the corresponding slope and offset values $s = 4.50681$ and $d = 0.0991499$, respectively (Eqn. (4.34)). The resulting correction function $\bar{f}_{\text{ITU}}(a)$ has a *nominal* gamma value of 0.45, which corresponds to the *effective* gamma value $\gamma_{\text{eff}} = 1/1.956 \approx 0.511$. The gamma correction in the sRGB standard [40] is specified on the same basis (with different parameters; see Vol. 2 [6, Sec. 6.3]).

Figure 4.22 shows the actual correction functions for the ITU and sRGB standards, respectively, each in comparison with the equivalent ordinary gamma correction. The ITU function (Fig. 4.22 (a)) with $\gamma = 0.45$ and $a_0 = 0.018$ corresponds to an ordinary gamma correction with effective gamma value $\gamma_{\text{eff}} = 0.511$ (dashed line). The curves for sRGB (Fig. 4.22 (b)) differ only by the parameters γ and a_0, as summarized in Table 4.1.

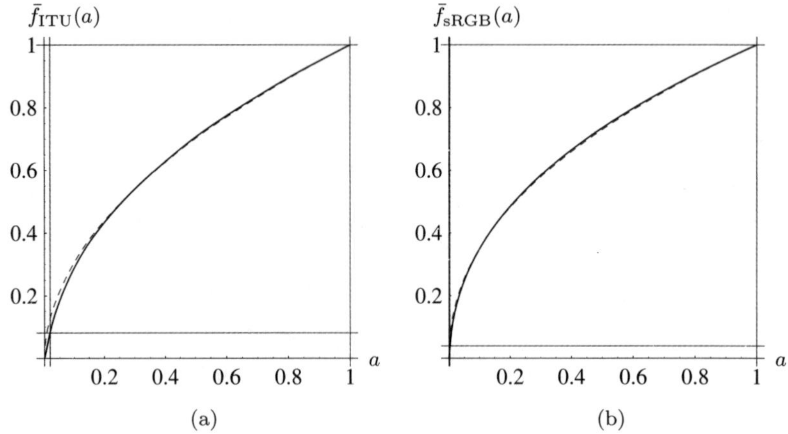

Figure 4.22 Gamma correction functions specified by the ITU-R BT.709 (a) and sRGB (b) standards. The continuous plot shows the modified gamma correction with the nominal gamma value γ and transition point a_0. The dashed lines mark the equivalent ordinary gamma correction with effective gamma γ_{eff}.

Inverting the modified gamma correction

To invert the modified gamma correction of the form $b = \bar{f}_{\gamma,a_0}(a)$ (Eqn. (4.33)), we need the inverse of the function $\bar{f}_{\gamma,a_0}()$, which is again defined in two parts,

$$\bar{f}^{-1}_{\gamma,a_0}(b) = \begin{cases} b/s & \text{for } 0 \leq b \leq s \cdot a_0 \\ \left(\dfrac{b+d}{1+d}\right)^{1/\gamma} & \text{for } s \cdot a_0 < b \leq 1, \end{cases} \tag{4.35}$$

where s and d are the values defined in Eqn. (4.34). The inverse gamma correction function is required in particular for transforming between different color spaces if nonlinear (i. e., gamma-corrected) component values are involved (see also Vol. 2 [6, Sec. 6.2]).

4.8 Point Operations in ImageJ

Several important types of point operations are already implemented in ImageJ, so there is no need to program every operation manually (as shown in Prog. 4.4). In particular, it is possible in ImageJ to apply point operations efficiently by using tabulated functions, to use built-in standard functions for point operations on single images, and to apply arithmetic operations on pairs of images. These issues are described briefly in the remaining parts of this section.

4.8.1 Point Operations with Lookup Tables

Some point operations require complex computations for each pixel, and the processing of large images may be quite time-consuming. If the point operation is *homogeneous* (i. e., independent of the pixel coordinates), the value of the mapping function can be precomputed for every possible pixel value and stored in a lookup table, which may then be applied very efficiently to the image. A lookup table **L** represents a discrete mapping (function f) from the original to the new pixel values,

$$\mathbf{L} : [0, K-1] \xrightarrow{f} [0, K-1]. \tag{4.36}$$

For a point operation specified by a particular pixel mapping function $a' = f(a)$, the table **L** is initialized with the values

$$\mathbf{L}(a) \leftarrow f(a) \qquad \text{for } 0 \leq a < K. \tag{4.37}$$

Thus the K table elements of **L** need only be computed once, where typically $K = 256$. Performing the actual point operation only requires a simple (and quick) table lookup in **L** at each pixel,

$$I(u, v) \leftarrow \mathbf{L}(I(u, v)), \tag{4.38}$$

which is much more efficient than any individual function call. ImageJ provides the method

```
void applyTable(int[] lut)
```

for objects of type `ImageProcessor`, which requires a lookup table *`lut`* (**L**) as a one-dimensional `int` array of size K (see Prog. 4.4 on page 83 for an example). The advantage of this approach is obvious: for an 8-bit image, for example, the mapping function is evaluated only 256 times (independent of the image size) and not a million times or more as in the case of a large image. The use of lookup tables for implementing point operations thus always makes sense if the number of image pixels ($M \times N$) is greater than the number of possible pixel values K (which is usually the case).

4.8.2 Arithmetic Operations

ImageJ implements a set of common arithmetic operations as methods for the class `ImageProcessor`, which are summarized in Table 4.2. In the following example, the image is multiplied by a scalar constant (1.5) to increase its contrast:

```
ImageProcessor ip = ... //some image ip
ip.multiply(1.5);
```

Table 4.2 ImageJ methods for arithmetic operations applicable to objects of type `ImageProcessor`.

Method	Operation
`void abs()`	$I(u,v) \leftarrow \lvert I(u,v) \rvert$
`void add(int p)`	$I(u,v) \leftarrow I(u,v) + p$
`void gamma(double g)`	$I(u,v) \leftarrow \big(I(u,v)/255\big)^g \cdot 255$
`void invert(int p)`	$I(u,v) \leftarrow 255 - I(u,v)$
`void log()`	$I(u,v) \leftarrow \log_{10}\big(I(u,v)\big)$
`void max(double s)`	$I(u,v) \leftarrow \max\big(I(u,v), s\big)$
`void min(double s)`	$I(u,v) \leftarrow \min\big(I(u,v), s\big)$
`void multiply(double s)`	$I(u,v) \leftarrow \mathrm{round}\big(I(u,v) \cdot s\big)$
`void sqr()`	$I(u,v) \leftarrow I(u,v)^2$
`void sqrt()`	$I(u,v) \leftarrow \sqrt{I(u,v)}$

The image `ip` is destructively modified by all of these methods, with the results being limited (clamped) to the minimum and maximum pixel values, respectively.

4.8.3 Point Operations Involving Multiple Images

Point operations may involve more than one image at once, with arithmetic operations on the pixels of *pairs* of images being a special but important case. For example, we can express the pointwise *addition* of two images I_1 and I_2 (of identical size) to create a new image I' as

$$I'(u,v) \;\leftarrow\; I_1(u,v) + I_2(u,v) \tag{4.39}$$

for all positions (u,v). In general, any function $f(a_1, a_2, \ldots, a_n)$ over n pixel values a_i may be defined to perform pointwise combinations of n images, i. e.,

$$I'(u,v) \;\leftarrow\; f\big(I_1(u,v), I_2(u,v), \ldots I_n(u,v)\big). \tag{4.40}$$

However, most arithmetic operations on multiple images required in practice can be implemented as sequences of successive binary operations on pairs of images.

4.8.4 Methods for Point Operations on Two Images

ImageJ supplies a single method for implementing arithmetic operations on pairs of images,

Table 4.3 Arithmetic operations for pairs of images provided by `ImageProcessor`'s `copyBits()` method. The constants `ADD`, etc., listed in this table are defined by ImageJ's `Blitter` interface.

`ADD`	$I_1(u,v) \leftarrow I_1(u,v) + I_2(u,v)$
`AVERAGE`	$I_1(u,v) \leftarrow \big(I_1(u,v) + I_2(u,v)\big) \,/\, 2$
`DIFFERENCE`	$I_1(u,v) \leftarrow \lvert I_1(u,v) - I_2(u,v)\rvert$
`DIVIDE`	$I_1(u,v) \leftarrow I_1(u,v) \,/\, I_2(u,v)$
`MAX`	$I_1(u,v) \leftarrow \max\big(I_1(u,v), I_2(u,v)\big)$
`MIN`	$I_1(u,v) \leftarrow \min\big(I_1(u,v), I_2(u,v)\big)$
`MULTIPLY`	$I_1(u,v) \leftarrow I_1(u,v) \cdot I_2(u,v)$
`SUBTRACT`	$I_1(u,v) \leftarrow I_1(u,v) - I_2(u,v)$

```
void copyBits(ImageProcessor ip2, int u, int v, int mode),
```

which applies the binary operation specified by the transfer mode parameter `mode` to all pixel pairs taken from the *source image* `ip2` and the *target image* (the image on which this method is invoked) and stores the result in the target image. `u`, `v` are the coordinates where the source image is inserted into the target image (usually `u` = `v` = 0). The code fragment in the following example demonstrates the addition of two images:

```
// add images ip1 and ip2:
ImageProcessor ip1 = ... // target image I1
ImageProcessor ip2 = ... // source image I2
...
ip1.copyBits(ip2, 0, 0, Blitter.ADD); // I1(u,v) <- I1(u,v) + I2(u,v)
// ip1 holds the result, ip2 is unchanged
...
```

By this operation, the target image `ip1` is destructively modified, while the source image `ip2` remains unchanged. The constant `ADD` is one of several arithmetic transfer modes defined by the `Blitter` interface (see Table 4.3). In addition, `Blitter` defines (bitwise) logical operations, such as `OR` and `AND`.[9] For arithmetic operations, the `copyBits()` method limits the results to the admissible range of pixel values (of the target image). Also note that (except for target images of type `FloatProcessor`) the results are *not* rounded but truncated to integer values.

[9] See also Sec. 10 of the ImageJ Short Reference [5].

4.8.5 ImageJ Plugins Involving Multiple Images

ImageJ provides two types of plugin: a generic plugin (`PlugIn`), which can be run without any open image, and plugins of type `PlugInFilter`, which apply to a single image. In the latter case, the currently active image is passed as an object of type `ImageProcessor` to the plugin's `run()` method (see also Sec. 2.2.3).

If two or more images $I_1, I_2 \ldots I_k$ are to be combined by a plugin program, only a single image I_1 can be passed directly to the plugin's `run()` method, but not the additional images $I_2 \ldots I_k$. The usual solution is to make the plugin open a dialog window to let the user select the remaining images interactively. This is demonstrated in the following example plugin for transparently blending two images.

Example: alpha blending

Alpha blending is a simple method for transparently overlaying two images, I_{BG} and I_{FG}. The background image I_{BG} is covered by the foreground image I_{FG}, whose transparency is controlled by the value α in the form

$$I'(u,v) \leftarrow \alpha \cdot I_{\mathrm{BG}}(u,v) + (1-\alpha) \cdot I_{\mathrm{FG}}(u,v) \tag{4.41}$$

with $0 \leq \alpha \leq 1$. For $\alpha = 0$, the foreground image I_{FG} is nontransparent (opaque) and thus entirely hides the background image I_{BG}. Conversely, the image I_{FG} is fully transparent for $\alpha = 1$ and only I_{BG} is visible. All α values between 0 and 1 result in a weighted sum of the corresponding pixel values taken from I_{BG} and I_{FG} (Eqn. (4.41)).

Figure 4.23 shows the results of alpha blending for different α values. The Java code for the corresponding implementation (as an ImageJ plugin) is listed in Progs. 4.5 and 4.6. The background image (`bgIp`) is passed directly to the plugin's `run()` method. The second (foreground) image and the α value are specified interactively by creating an instance of the ImageJ class `GenericDialog`, which allows the simple implementation of dialog windows with various types of input fields.[10]

[10] See also Sec. 18.2 of the ImageJ Short Reference [5], where a similar example producing a *stack* of images by stepwise alpha blending can be found in Sec. 15.3.

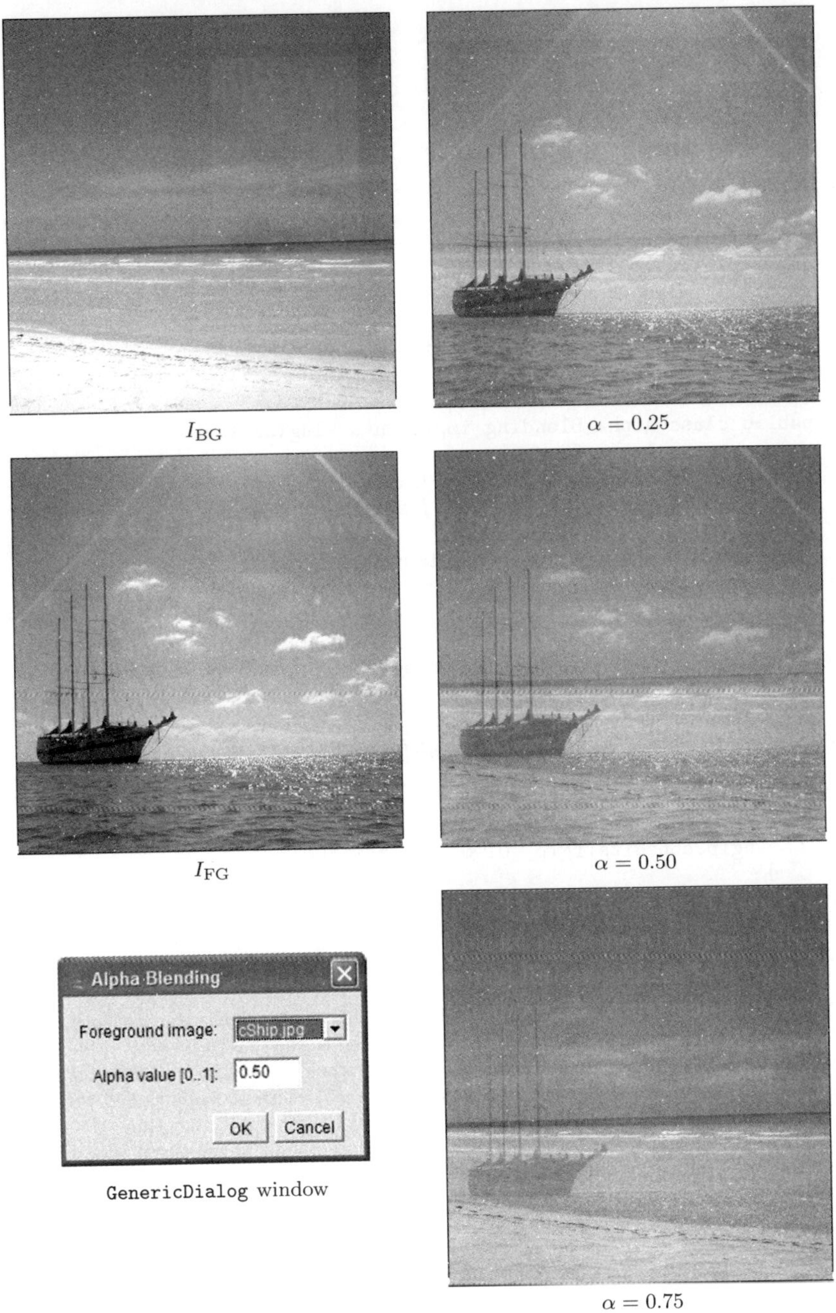

Figure 4.23 Alpha blending example. Background image (I_{BG}) and foreground image (I_{FG}), `GenericDialog` window (see the implementation in Progs. 4.5 and 4.6), and blended images for transparency values $\alpha = 0.25, 0.50,$ and 0.75.

```
1 import ij.IJ;
2 import ij.ImagePlus;
3 import ij.WindowManager;
4 import ij.gui.GenericDialog;
5 import ij.plugin.filter.PlugInFilter;
6 import ij.process.*;
7
8 public class Alpha_Blending implements PlugInFilter {
9
10   static double alpha = 0.5; // transparency of foreground image
11   ImagePlus fgIm = null;     // foreground image
12
13   public int setup(String arg, ImagePlus imp) {
14     return DOES_8G;
15   }
16
17   public void run(ImageProcessor bgIp) { // background image
18     if(runDialog()) {
19       ImageProcessor fgIp
20         = fgIm.getProcessor().convertToByte(false);
21       fgIp = fgIp.duplicate();
22       fgIp.multiply(1-alpha);
23       bgIp.multiply(alpha);
24       bgIp.copyBits(fgIp, 0, 0, Blitter.ADD);
25     }
26   }
27
28   // continued ...
```

Program 4.5 Alpha blending plugin (part 1). A background image is transparently blended with a selected foreground image. The plugin is applied to the (currently active) background image, and the foreground image must also be open when the plugin is started. The background image (`bgIp`), which is passed to the plugin's `run()` method, is multiplied with α (line 23). The foreground image (`fgIP`, selected in part 2) is first duplicated (line 21) and then multiplied with $(1-\alpha)$ (line 22). Thus the original foreground image is not modified. The final result is obtained by adding the two weighted images (line 24).

```
30  // class Alpha_Blending (continued)
31
32  boolean runDialog() {
33    // get list of open images
34    int[] windowList = WindowManager.getIDList();
35    if (windowList == null){
36      IJ.noImage();
37      return false;
38    }
39
40    // get all image titles
41    String[] windowTitles = new String[windowList.length];
42    for (int i = 0; i < windowList.length; i++) {
43      ImagePlus im = WindowManager.getImage(windowList[i]);
44      if (im == null)
45        windowTitles[i] = "untitled";
46      else
47        windowTitles[i] = im.getShortTitle();
48    }
49
50    // create dialog and show
51    GenericDialog gd = new GenericDialog("Alpha Blending");
52    gd.addChoice("Foreground image:",
53        windowTitles, windowTitles[0]);
54    gd.addNumericField("Alpha value [0..1]:", alpha, 2);
55    gd.showDialog();
56    if (gd.wasCanceled())
57      return false;
58    else {
59      int fgIdx = gd.getNextChoiceIndex();
60      fgIm = WindowManager.getImage(windowList[fgIdx]);
61      alpha = gd.getNextNumber();
62      return true;
63    }
64  }
65
66 } // end of class Alpha_Blending
```

Program 4.6 Alpha blending plugin (part 2). To select the foreground image, a list of currently open images and image titles is obtained (lines 34–42). Then a dialog object (`GenericDialog`) is created and opened for specifying the foreground image (`fgIm`) and the α value (`alpha`). `fgIm` and `alpha` are variables in the class `AlphaBlend_` (declared in part 1, Prog. 4.5). The `runDialog()` method returns `true` if successful and `false` if no images are open or the dialog was canceled by the user.

4.9 Exercises

Exercise 4.1

Implement the auto-contrast operation as defined in Eqns. (4.8)–(4.10) as an ImageJ plugin for an 8-bit grayscale image. Set the quantile q of pixels to be saturated at both ends of the intensity range (0 and 255) to $q = q_{\mathrm{low}} = q_{\mathrm{high}} = 1\%$.

Exercise 4.2

Modify the histogram equalization plugin in Prog. 4.2 to use a lookup table (Sec. 4.8.1) for computing the point operation.

Exercise 4.3

Implement the histogram equalization as defined in Eqn. (4.11), but use the *modified* cumulative histogram defined in Eqn. (4.12), cumulating the square root of the histogram entries. Compare the results to the standard (linear) approach by plotting the resulting histograms and cumulative histograms as shown in Fig. 4.10.

Exercise 4.4

Show formally that (a) a linear histogram equalization (Eqn. (4.11)) does not change an image that already has a uniform intensity distribution and (b) that any repeated application of histogram equalization to the same image causes no more changes.

Exercise 4.5

Show that the linear histogram equalization (Sec. 4.5) is only a special case of histogram specification (Sec. 4.6).

Exercise 4.6

Implement (in Java) a histogram specification using a piecewise linear reference distribution function, as described in Sec. 4.6.3. Define a new object class with all necessary instance variables to represent the distribution function and implement the required functions $\mathsf{P}_L(i)$ (Eqn. (4.22)) and $\mathsf{P}_L^{-1}(b)$ (Eqn. (4.23)) as methods of this class.

Exercise 4.7

Using a histogram specification for adjusting *multiple* images (Sec. 4.6.4), one could either use one typical image as the reference or compute an "average" reference histogram from a set of images. Implement the second approach and discuss its possible advantages (or disadvantages).

Exercise 4.8

Implement the modified gamma correction (Eqn. (4.33)) as an ImageJ plugin with variable values for γ and a_0 using a lookup table as shown in Prog. 4.4.

Exercise 4.9

Show that the modified gamma correction function $\bar{f}_{\gamma,a_0}(a)$, with the parameters defined in Eqns. (4.33) and (4.34), is C1-continuous (i. e., both the function itself and its first derivative are continuous).

5
Filters

The essential property of point operations (discussed in the previous chapter) is that each new pixel value only depends on the original pixel at the *same* position. The capabilities of point operations are limited, however. For example, they cannot accomplish the task of *sharpening* or *smoothing* an image (Fig. 5.1). This is what filters can do. They are similar to point operations in the sense that they also produce a 1:1 mapping of the image coordinates, i. e., the geometry of the image does not change.

5.1 What Is a Filter?

The main difference between filters and point operations is that filters generally use more than one pixel from the source image for computing each new pixel value. Let us first take a closer look at the task of smoothing an image. Images look sharp primarily at places where the local intensity rises or drops sharply (i. e., where the difference between neighboring pixels is large). On the other hand, we perceive an image as blurred or fuzzy where the local intensity function is smooth.

A first idea for smoothing an image could thus be to simply replace every pixel by the *average* of its neighboring pixels. To determine the new pixel value in the smoothed image $I'(u,v)$, we use the original pixel $I(u,v) = p_0$ at the same position plus its eight neighboring pixels $p_1, p_2, \ldots p_8$ to compute the arithmetic mean of these nine values,

$$I'(u,v) \leftarrow \frac{p_0 + p_1 + p_2 + p_3 + p_4 + p_5 + p_6 + p_7 + p_8}{9} \tag{5.1}$$

W. Burger, M.J. Burge, *Principles of Digital Image Processing*, Undergraduate Topics in Computer Science, DOI 10.1007/978-1-84800-191-6_5, © Springer-Verlag London Limited, 2009

Figure 5.1 No point operation can blur or sharpen an image. This is an example of what filters can do. Like point operations, filters do not modify the geometry of an image.

or, expressed in relative image coordinates,

$$\begin{aligned} I'(u,v) \leftarrow \tfrac{1}{9} \cdot [\, & I(u-1,v-1) + I(u,v-1) + I(u+1,v-1) + \\ & I(u-1,v) + I(u,v) + I(u+1,v) + \\ & I(u-1,v+1) + I(u,v+1) + I(u+1,v+1)\,]\,. \end{aligned} \tag{5.2}$$

Written more compactly, this is equivalent to

$$I'(u,v) \leftarrow \frac{1}{9} \cdot \sum_{j=-1}^{1} \sum_{i=-1}^{1} I(u+i, v+j). \tag{5.3}$$

This simple local averaging already exhibits all the important elements of a typical filter. In particular, it is a so-called *linear* filter, which is a very important class of filters. But how are filters defined in general? First they differ from point operations mainly by using not a single source pixel but a *set* of them for computing each resulting pixel. The coordinates of the source pixels are fixed relative to the current image position (u, v) and usually form a contiguous region, as illustrated in Fig. 5.2.

The *size* of the filter region is an important parameter of the filter because it specifies how many original pixels contribute to each resulting pixel value and thus determines the spatial extent (support) of the filter. For example, the smoothing filter in Eqn. (5.2) uses a 3×3 region of support that is centered at the current coordinate (u, v). Similar filters with larger support, such as 5×5, 7×7, or even 21×21 pixels, would obviously have stronger smoothing effects.

The *shape* of the filter region is not necessarily quadratic or even rectangular. In fact, a circular (disk-shaped) region would be preferred to obtain an *isotropic* blur effect (i.e., one that is the same in all image directions). Another option is to assign different *weights* to the pixels in the support region, such as to give stronger emphasis to pixels that are closer to the center of the region. Furthermore, the support region of a filter does not need to be contiguous

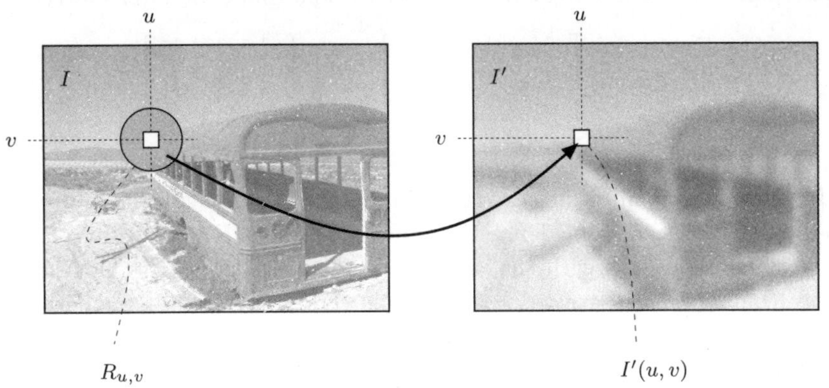

Figure 5.2 Principal filter operation. Each new pixel value $I'(u,v)$ is computed as a function of the pixels in a corresponding region of source pixels $R_{u,v}$ in the original image I.

and may not even contain the original pixel itself (imagine a ring-shaped filter region, for example).

It is probably confusing to have so many options—a more systematic method is needed for specifying and applying filters in a targeted manner. The traditional and proven classification into *linear* and *nonlinear* filters is based on the mathematical properties of the filter function; i. e., whether the result is computed from the source pixels by a *linear* or a *nonlinear* expression. In the following, we discuss both classes of filters and show several practical examples.

5.2 Linear Filters

Linear filters are denoted that way because they combine the pixel values in the support region in a linear fashion; i. e., as a weighted summation. The local averaging process discussed in the beginning (Eqn. (5.3)) is a special example, where all nine pixels in the 3×3 support region are added with identical weights ($1/9$). With the same mechanism, a multitude of filters with different properties can be defined by simply modifying the distribution of the individual weights.

5.2.1 The Filter Matrix

For any linear filter, the size and shape of the support region, as well as the individual pixel weights, are specified by the "filter matrix" or "filter mask" $H(i,j)$. The size of the matrix H equals the size of the filter region, and every element $H(i,j)$ specifies the weight of the corresponding pixel in the summation. For the 3×3 smoothing filter in Eqn. (5.3), the filter matrix is

$$H(i,j) = \begin{bmatrix} 1/9 & 1/9 & 1/9 \\ 1/9 & 1/9 & 1/9 \\ 1/9 & 1/9 & 1/9 \end{bmatrix} = \frac{1}{9} \begin{bmatrix} 1 & 1 & 1 \\ 1 & 1 & 1 \\ 1 & 1 & 1 \end{bmatrix} \quad (5.4)$$

because each of the nine pixels contributes one-ninth of its value to the result.

In principle, the filter matrix $H(i,j)$ is, just like the image itself, a discrete, two-dimensional, real-valued function, $H : \mathbb{Z} \times \mathbb{Z} \mapsto \mathbb{R}$. The filter has its own coordinate system with the origin—often referred to as the "hot spot"—mostly (but not necessarily) located at the center. Thus, filter coordinates are generally positive and negative (Fig. 5.3). The filter function is of infinite extent and considered zero outside the region defined by the matrix H.

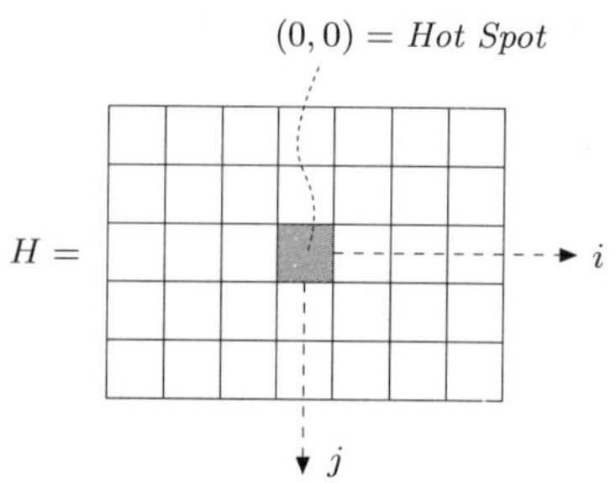

Figure 5.3 Filter matrix and coordinate system. i is the horizontal (column) index, j is the vertical (row) index.

5.2.2 Applying the Filter

For a linear filter, the result is unambiguously and completely specified by the coefficients of the filter matrix. Applying the filter to an image is a simple process that is illustrated in Fig. 5.4. The following steps are performed at each image position (u, v):

1. The filter matrix H is moved over the original image I such that its origin $H(0,0)$ coincides with the current image position (u, v).
2. All filter coefficients $H(i,j)$ are multiplied with the corresponding image element $I(u+i, v+j)$, and the results are added.
3. Finally, the resulting sum is stored at the current position in the new image $I'(u, v)$.

Described formally, pixels in the new image $I'(u,v)$ are computed by the operation

$$I'(u,v) \leftarrow \sum_{(i,j) \in R_H} I(u+i, v+j) \cdot H(i,j), \quad (5.5)$$

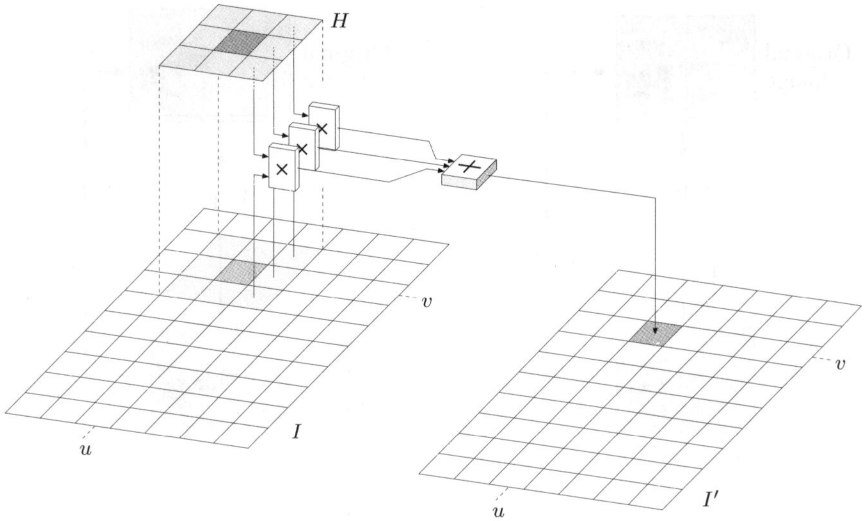

Figure 5.4 Linear filter. The filter matrix H is placed with its origin at position (u, v) on the image I. Each filter coefficient $H(i, j)$ is multiplied with the corresponding image pixel $I(u+i, v+j)$, the results are added, and the final sum is inserted as the new pixel value $I'(u, v)$.

where R_H denotes the set of coordinates covered by the filter H. For a typical 3×3 filter with centered origin, this is

$$I'(u,v) \leftarrow \sum_{i=-1}^{i=1} \sum_{j=-1}^{j=1} I(u+i, v+j) \cdot H(i,j), \tag{5.6}$$

for all image coordinates (u, v). Not quite for *all* coordinates, to be exact. There is an obvious problem at the image borders where the filter reaches outside the image and finds no corresponding pixel values to use in computing a result. For the moment, we ignore this border problem, but we will attend to it again in Sec. 5.5.2.

5.2.3 Computing the Filter Operation

Now that we understand the principal operation of a filter (Fig. 5.4) and know that the borders need special attention, we go ahead and program a simple linear filter in ImageJ. But before we do this, we may want to consider one more detail. In a point operation (e. g., in Progs. 4.1 and 4.2), each new pixel value depends only on the corresponding pixel value in the original image, and it was thus no problem simply to store the results back to the same image—the computation is done "in place" without the need for any intermediate storage.

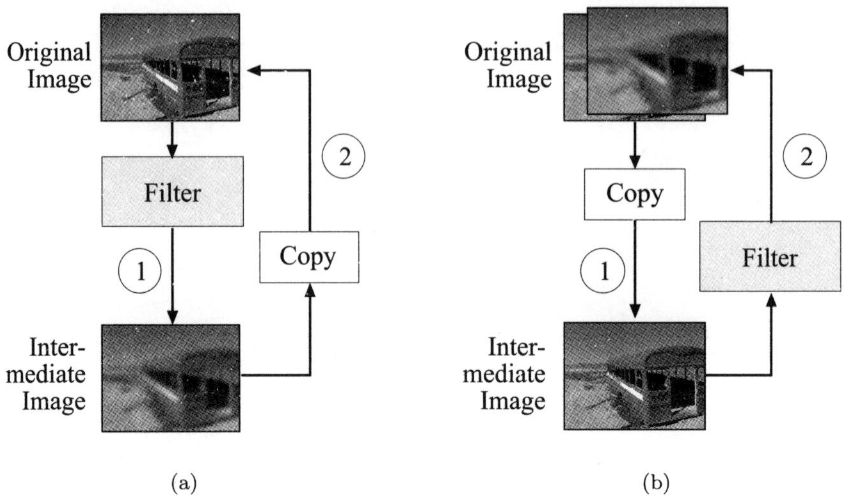

Figure 5.5 Practical implementation of filter operations. **Version A** (a): The result of the filter is first stored in an intermediate image and subsequently copied back to the original image. **Version B** (b): The original image is first copied to an intermediate image that serves as the source for the actual filter operation. The result replaces the pixels in the original image.

In-place computation is generally not possible for a filter since any original pixel contributes to more than one resulting pixel and thus may not be modified before all operations are complete. We therefore require additional storage space for the resulting image, which subsequently could be copied back to the source image again (if desired). Thus the complete filter operation can be implemented in two different ways (Fig. 5.5):

A. The result of the filter computation is initially stored in a new image whose content is eventually copied back to the original image.

B. The original image is first copied to an intermediate image that serves as the source for the filter computation. The results are directly stored in the original image.

The same amount of storage is required for both versions, and thus none of them offers a particular advantage. In the following examples, we generally use version B.

5.2.4 Filter Plugin Examples

The following examples demonstrate the implementation of two very basic filters that are nevertheless often used in practice.

```
1 import ij.*;
2 import ij.plugin.filter.PlugInFilter;
3 import ij.process.*;
4
5 public class Filter_Average3x3 implements PlugInFilter {
6     ...
7     public void run(ImageProcessor orig) {
8         int w = orig.getWidth();
9         int h = orig.getHeight();
10        ImageProcessor copy = orig.duplicate();
11
12        for (int v = 1; v <= h-2; v++) {
13            for (int u = 1; u <= w-2; u++) {
14                //compute filter result for position (u,v)
15                int sum = 0;
16                for (int j = -1; j <= 1; j++) {
17                    for (int i = -1; i <= 1; i++) {
18                        int p = copy.getPixel(u+i, v+j);
19                        sum = sum + p;
20                    }
21                }
22                int q = (int) Math.round(sum/9.0);
23                orig.putPixel(u, v, q);
24            }
25        }
26    }
27 } // end of class Filter_Average3x3
```

Program 5.1 3×3 averaging "box" filter (ImageJ plugin). First (in line 10) a duplicate (`copy`) of the original image (`orig`) is created, which is used as the source image in the subsequent filter computation (line 18). In line 22, the result for the current image position (`u, v`) is rounded and subsequently stored in the original image (line 23). Notice that the border pixels remain unchanged because they are not reached by the iteration over (`u, v`).

Simple 3×3 averaging filter ("box" filter)

Program 5.1 shows the ImageJ code for a simple 3×3 smoothing filter based on local averaging (Eqn. (5.4)), which is often called a "box" filter because of its box-like shape. No explicit filter matrix is required in this case since all filter coefficients are identical ($1/9$). Also, no *clamping* (see Sec. 4.1.2) of the results is needed because the sum of the filter coefficients is 1 and thus no pixel values outside the admissible range can be created.

Although this example implements an extremely simple filter, it nevertheless demonstrates the general structure of a two-dimensional filter program. In particular, *four* nested loops are needed: *two* (outer) loops for moving the filter over the image coordinates (u, v) and *two* (inner) loops to iterate over the (i, j) coordinates within the rectangular filter region. The required amount of computation thus depends not only upon the size of the image but equally on the size of the filter.

Another 3 × 3 smoothing filter

Instead of the constant weights applied in the previous example, we now use a real filter matrix with variable coefficients. For this purpose, we apply a bell-shaped 3×3 filter function $H(i, j)$, which puts more emphasis on the center pixel than the surrounding pixels:

$$H(i,j) = \begin{bmatrix} 0.075 & 0.125 & 0.075 \\ 0.125 & \mathbf{0.200} & 0.125 \\ 0.075 & 0.125 & 0.075 \end{bmatrix}. \tag{5.7}$$

Notice that all coefficients in H are positive and sum to 1 (i. e., the matrix is normalized) such that all results remain within the original range of pixel values. Again no clamping is necessary and the program structure in Prog. 5.2 is virtually identical to the previous example. The filter matrix (`filter`) is represented by a two-dimensional array[1] of type `double`. Each pixel is multiplied by the corresponding coefficient of the filter matrix, the resulting sum being also of type `double`. Accessing the filter coefficients, it must be considered that the coordinate origin of the filter matrix is assumed to be at its center (i. e., at position $(1, 1)$) in the case of a 3×3 matrix. This explains the offset of 1 for the i and j coordinates (see Prog. 5.2, line 20).

5.2.5 Integer Coefficients

Instead of using floating-point coefficients (as in the previous examples), it is often simpler and usually more efficient to work with integer coefficients in combination with some common scale factor s,

$$H(i,j) = s \cdot H'(i,j), \tag{5.8}$$

with $H'(i, j) \in \mathbb{Z}$ and $s \in \mathbb{R}$. If all filter coefficients are positive (which is the case for any smoothing filter), then s is usually taken as the reciprocal of the sum of the coefficients,

$$s = \frac{1}{\sum_{i,j} H'(i,j)}, \tag{5.9}$$

to obtain a normalized filter matrix. In this case, the results are bounded to the original range of pixel values. For example, the filter matrix in Eqn. (5.7) could be defined equivalently as

$$H(i,j) = \begin{bmatrix} 0.075 & 0.125 & 0.075 \\ 0.125 & \underline{0.200} & 0.125 \\ 0.075 & 0.125 & 0.075 \end{bmatrix} = \frac{1}{40} \begin{bmatrix} 3 & 5 & 3 \\ 5 & \underline{8} & 5 \\ 3 & 5 & 3 \end{bmatrix}, \tag{5.10}$$

[1] See the additional comments in Appendix B.2.4 regarding two-dimensional arrays in Java.

```
1  public void run(ImageProcessor orig) {
2      int w = orig.getWidth();
3      int h = orig.getHeight();
4      // 3 × 3 filter matrix
5      double[][] filter = {
6          {0.075, 0.125, 0.075},
7          {0.125, 0.200, 0.125},
8          {0.075, 0.125, 0.075}
9      };
10     ImageProcessor copy = orig.duplicate();
11
12     for (int v = 1; v <= h-2; v++) {
13         for (int u = 1; u <= w-2; u++) {
14             // compute filter result for position (u, v)
15             double sum = 0;
16             for (int j = -1; j <= 1; j++) {
17                 for (int i = -1; i <= 1; i++) {
18                     int p = copy.getPixel(u+i, v+j);
19                     // get the corresponding filter coefficient:
20                     double c = filter[j+1][i+1];
21                     sum = sum + c * p;
22                 }
23             }
24             int q = (int) Math.round(sum);
25             orig.putPixel(u, v, q);
26         }
27     }
28 }
```

Program 5.2 3×3 smoothing filter (ImageJ plugin, run() method only). The filter matrix is defined as a two-dimensional array of type `double` (line 5). The coordinate origin of the filter is assumed to be at the center of the matrix (i.e., at the array position $[1,1]$), which is accounted for by an offset of 1 for the i, j coordinates in line 20. The results are rounded (line 24) and stored in the original image (line 25).

with the common scale factor $s = \frac{1}{40} = 0.025$. A similar scaling is used in Prog. 5.3.

In Adobe Photoshop, linear filters can be specified with the "Custom Filter" tool (Fig. 5.6) using integer coefficients and a common scale factor Scale (which corresponds to the reciprocal of s). In addition, a constant Offset value can be specified; e.g., to shift negative results (caused by negative coefficients) into the visible range of values. In summary, the operation performed by the 5×5 Photoshop custom filter can be expressed as

$$I'(u,v) \leftarrow \mathsf{Offset} + \frac{1}{\mathsf{Scale}} \sum_{j=-2}^{j=2} \sum_{i=-2}^{i=2} I(u+i, v+j) \cdot H(i,j). \tag{5.11}$$

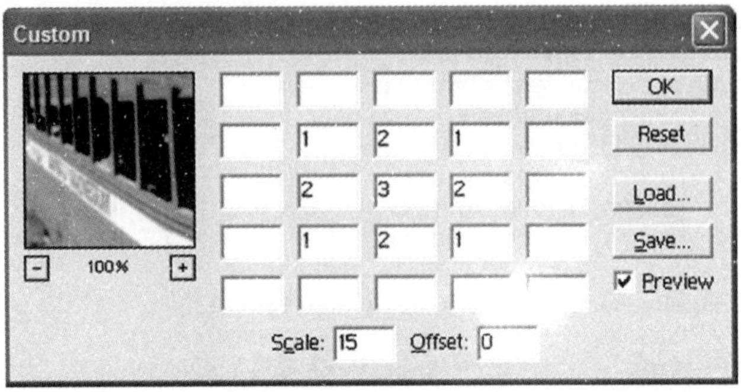

Figure 5.6 Adobe Photoshop's "Custom Filter" implements linear filters up to a size of 5×5. The filter's coordinate origin ("hot spot") is assumed to be at the center (value set to 3 in this example), and empty cells correspond to zero coefficients. In addition to the (integer) coefficients, common Scale and Offset values can be specified (see Eqn. (5.11)).

5.2.6 Filters of Arbitrary Size

Small filters of size 3×3 are frequently used in practice, but sometimes much larger filters are required. Let us assume that the filter matrix is centered and has an odd number of $(2K+1)$ columns and $(2L+1)$ rows, with $K, L \geq 0$. If the image is of size $M \times N$, that is

$$I(u, v) \quad \text{with} \quad 0 \leq u < M \quad \text{and} \quad 0 \leq v < N,$$

then the filter can be computed for all image coordinates (u', v') with

$$K \leq u' \leq (M-K-1) \qquad \text{and} \qquad L \leq v' \leq (N-L-1),$$

as illustrated in Fig. 5.7. Program 5.3 (which is adapted from Prog. 5.2) shows a 7×5 smoothing filter as an example for implementing linear filters of arbitrary size. This example uses integer-valued filter coefficients in combination with a common scale factor s, as described above. As usual, the "hot spot" of the filter is assumed to be at the matrix center, and the range of all iterations depends on the dimensions of the filter matrix. In this case, clamping of the results is included (in lines 33–34) as a preventive measure.

5.2.7 Types of Linear Filters

Since the effects of a linear filter are solely specified by the filter matrix (which can take on arbitrary values), an infinite number of different linear filters exists, at least in principle. So how can these filters be used and which filters are suited for a given task? In the following, we briefly discuss two broad classes of linear filters that are of key importance in practice: smoothing filters and difference filters (Fig. 5.8).

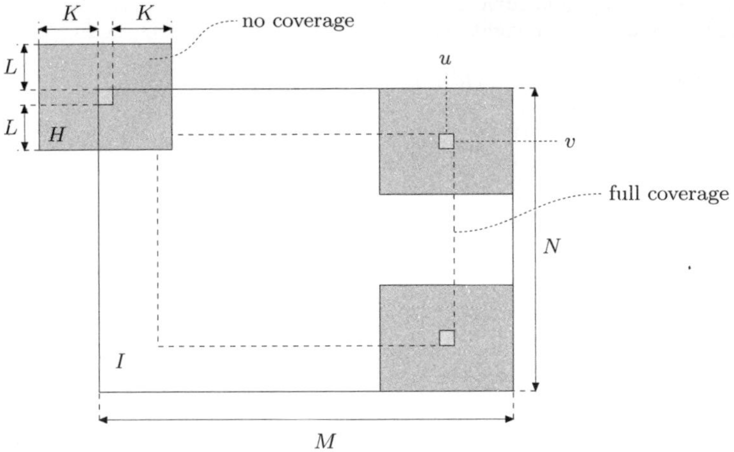

Figure 5.7 Border geometry. The filter can be applied only at locations (u, v) where the filter matrix H of size $(2K+1) \times (2L+1)$ is fully contained in the image (inner rectangle).

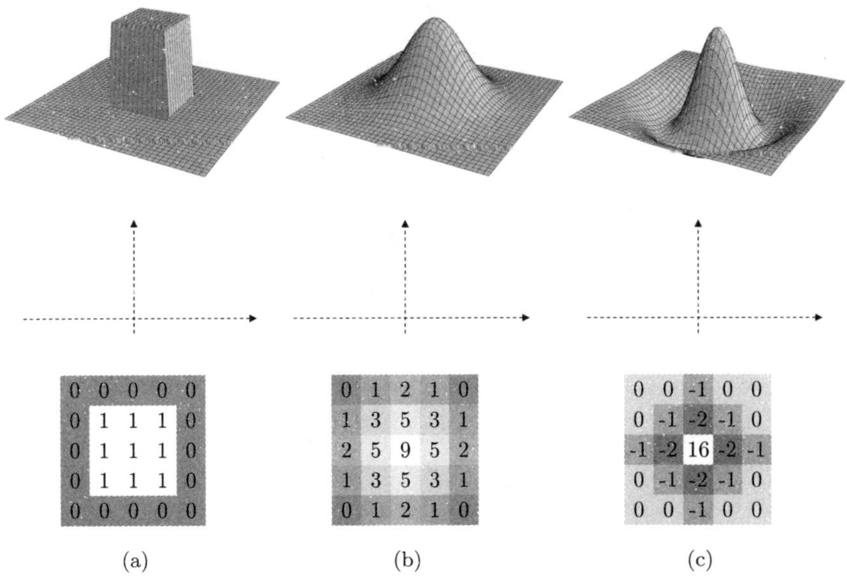

Figure 5.8 Typical examples of linear filters, illustrated as 3D plots (top), profiles (center), and approximations by discrete filter matrices (bottom). The "box" filter (a) and the Gauss filter (b) are both *smoothing filters* with all-positive coefficients. The "Laplace" or "Mexican hat" filter (c) is a *difference filter*. It computes the weighted difference between the center pixel and the surrounding pixels and thus reacts most strongly to local intensity peaks.

```
1  public void run(ImageProcessor orig) {
2      int M = orig.getWidth();
3      int N = orig.getHeight();
4
5      // filter matrix of size (2K + 1) × (2L + 1)
6      int[][] filter = {
7          {0,0,1,1,1,0,0},
8          {0,1,1,1,1,1,0},
9          {1,1,1,1,1,1,1},
10         {0,1,1,1,1,1,0},
11         {0,0,1,1,1,0,0}
12     };
13
14     double s = 1.0/23;  // sum of filter coefficients is 23
15
16     int K = filter[0].length/2;
17     int L = filter.length/2;
18
19     ImageProcessor copy = orig.duplicate();
20
21     for (int v = L; v <= N-L-1; v++) {
22         for (int u = K; u <= M-K-1; u++) {
23             // compute filter result for position (u, v)
24             int sum = 0;
25             for (int j = -L; j <= L; j++) {
26                 for (int i = -K; i <= K; i++) {
27                     int p = copy.getPixel(u+i, v+j);
28                     int c = filter[j+L][i+K];
29                     sum = sum + c * p;
30                 }
31             }
32             int q = (int) Math.round(s * sum);
33             if (q < 0)  q = 0;
34             if (q > 255) q = 255;
35             orig.putPixel(u, v, q);
36         }
37     }
38 }
```

Program 5.3 ImageJ plugin (`run()` method only) for filters of arbitrary size. The filter matrix is an integer array of size $(2K+1) \times (2L+1)$ with the origin at the center element. The summation variable `sum` is also defined as an integer (`int`), which is scaled by a constant factor `s` and rounded in line 32. The border pixels are not modified.

Smoothing filters

Every filter we have discussed so far caused some kind of smoothing. In fact, any linear filter with positive-only coefficients is a smoothing filter in a sense because such a filter computes merely a weighted average of the image pixels within a certain image region.

Box filter. This simplest of all smoothing filters, whose 3D shape resembles a box (Fig. 5.8 (a)), is a well-known friend already. Unfortunately, the box filter is far from an optimal smoothing filter due to its wild behavior in frequency space, which is caused by the sharp cutoff around its sides. Described in frequency terms, smoothing corresponds to low-pass filtering (i. e., effectively attenuating all signal components above a given cutoff frequency).[2] The box filter, however, produces strong "ringing" in frequency space and is therefore not considered a high-quality smoothing filter. It may also appear rather ad hoc to assign the same weight to all image pixels in the filter region. Instead, one would probably expect to have stronger emphasis given to pixels near the center of the filter than to the more distant ones. Furthermore, smoothing filters should possibly operate "isotropically" (i. e., uniformly in each direction), which is certainly not the case for the rectangular box filter.

Gaussian filter. The filter matrix (Fig. 5.8 (b)) of this smoothing filter corresponds to a two-dimensional Gaussian function,

$$G_\sigma(x,y) = e^{-\frac{r^2}{2\sigma^2}} = e^{-\frac{x^2+y^2}{2\sigma^2}}, \tag{5.12}$$

where σ denotes the width (standard deviation) of the bell-shaped function and r is the distance (radius) from the center. The pixel at the center receives the maximum weight (1.0, which is scaled to the integer value 9 in the matrix shown in Fig. 5.8 (b)), and the remaining coefficients drop off smoothly with increasing distance from the center. The Gaussian filter is isotropic if the discrete filter matrix is large enough for a sufficient approximation (at least 5×5). As a low-pass filter, the Gaussian is "well-behaved" in frequency space and thus clearly superior to the box filter. The two-dimensional Gaussian filter is separable into a pair of one-dimensional filters (see Sec. 5.3.3), which facilitates its efficient implementation.

Difference filters

If some of the filter coefficients are negative, the filter calculation can be interpreted as the difference of two sums: the weighted sum of all pixels with associated positive coefficients minus the weighted sum of pixels with negative coefficients in the filter region R_H, that is

$$\begin{aligned} I'(u,v) = &\sum_{(i,j)\in R_H^+} I(u+i,v+j) \cdot |H(i,j)| \\ &- \sum_{(i,j)\in R_H^-} I(u+i,v+j) \cdot |H(i,j)|, \end{aligned} \tag{5.13}$$

---

[2] More details on the image vs. frequency space and related concepts are covered in Chapters 7 and 8 of Vol. 2 [6].

where R_H^+ and R_H^- denote the partitions of the filter with positive coefficients $H(i,j) > 0$ and negative coefficients $H(i,j) < 0$, respectively. For example, the 5×5 Laplace filter in Fig. 5.8 (c) computes the difference between the center pixel (with weight 16) and the weighted sum of 12 surrounding pixels (with weights -1 or -2). The remaining 12 pixels have associated zero coefficients and are thus ignored in the computation.

While local intensity variations are *smoothed* by averaging, we can expect the exact contrary to happen when differences are taken: local intensity changes are *enhanced.* Important applications of difference filters thus include edge detection (Sec. 6.2) and image sharpening (Sec. 6.6).

5.3 Formal Properties of Linear Filters

In the previous sections, we have approached the concept of filters in a rather casual manner to quickly get a grasp of how filters are defined and used. While such a level of treatment may be sufficient for most practical purposes, the power of linear filters may not really be apparent yet considering the limited range of (simple) applications seen so far.

The real importance of linear filters (and perhaps their formal elegance) only becomes visible when taking a closer look at some of the underlying theoretical details. At this point, it may be surprising to the experienced reader that we have not mentioned the term "convolution" in this context yet. We make up for this in the remaining parts of this section.

5.3.1 Linear Convolution

The operation associated with a linear filter, as described in the previous section, is not an invention of digital image processing but has been known in mathematics for a long time. It is called *linear convolution*[3] and in general combines two functions of the same dimensionality, either continuous or discrete. For discrete, two-dimensional functions I and H, the convolution operation is defined as

$$I'(u,v) = \sum_{i=-\infty}^{\infty} \sum_{j=-\infty}^{\infty} I(u-i, v-j) \cdot H(i,j), \qquad (5.14)$$

or

$$I' = I * H \qquad (5.15)$$

for short, where $*$ denotes the convolution operator. This almost looks the same as Eqn. (5.5), with two differences: the range of the variables i, j in the

[3] Oddly enough the simple concept of convolution is often (though unjustly) feared as an intractable mystery.

summation and the negative signs in the coordinates of $I(u-i, v-j)$. The first point is easy to explain: Because the coefficients outside the filter matrix $H(i, j)$, also referred to as a filter *kernel*, are assumed to be zero, the positions outside the matrix are irrelevant in the summation. To resolve the coordinate issue, we modify Eqn. (5.14) by replacing the summation variables i, j to

$$\begin{aligned} I'(u,v) &= \sum_{(i,j)\in R_H} I(u{-}i, v{-}j) \cdot H(i,j) \\ &= \sum_{(i,j)\in R_H} I(u{+}i, v{+}j) \cdot H(-i,-j) \\ &= \sum_{(i,j)\in R_H} I(u{+}i, v{+}j) \cdot H^*(i,j). \end{aligned} \tag{5.16}$$

The result is identical to the linear filter in Eqn. (5.5), with the filter function $H^*(i,j) = H(-i,-j)$ being the horizontally and vertically *reflected* (i. e., rotated by 180°) function H. To be precise, the operation in Eqn. (5.5) actually defines the linear *correlation*, which is merely a convolution with a reflected filter matrix.[4]

Thus the mathematical concept underlying all linear filters is the convolution operation ($*$), and its results are completely and sufficiently specified by the convolution matrix (or kernel) H. To illustrate this relationship, the convolution is often pictured as a "black box" operation, as shown in Fig. 5.9.

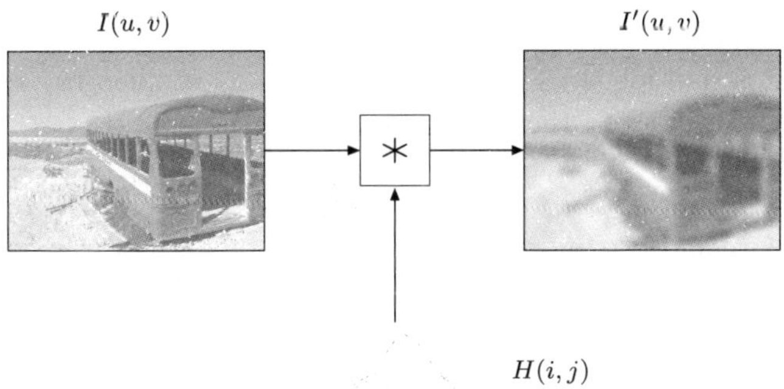

Figure 5.9 Convolution as a "black box" operation. The original image I is subjected to a linear convolution ($*$) with the convolution kernel H, producing the resulting image I'.

[4] Of course this is the same in the one-dimensional case. Linear correlation is typically used for comparing images or subpatterns (see Chapter 11 of Vol. 2 [6] for details).

5.3.2 Properties of Linear Convolution

The importance of linear convolution is based on its simple mathematical properties as well as its multitude of manifestations and applications. Linear convolution is a suitable model for many types of natural phenomena, including mechanical, acoustic, and optical systems. In particular (as shown in Ch. 7 of Vol. 2 [6]), there are strong formal links to the Fourier representation of signals in the frequency domain that are extremely valuable for understanding complex phenomena, such as sampling and aliasing. In the following, however, we first look at some important properties of linear convolution in the accustomed "signal" or image space.

Commutativity

Linear convolution is *commutative*; i. e.,

$$I * H = H * I. \tag{5.17}$$

Thus the result is the same if the image and filter kernel were interchanged, and it makes no difference if we convolve the image I with the kernel H or the other way around—the two functions I and H are exchangeable and may assume either role.

Linearity

Linear filters are called that way because of the linearity properties of the convolution operation, which manifests itself in various aspects. For example, if an image is multiplied by a scalar constant $s \in \mathbb{R}$, then the result of the convolution multiplies by the same factor,

$$(s \cdot I) * H \;=\; I * (s \cdot H) \;=\; s \cdot (I * H)\,. \tag{5.18}$$

Similarly, if we add two images I_1, I_2 pixel by pixel and convolve the resulting image with some kernel H, the same outcome is obtained by convolving each image individually and adding the two results afterward:

$$(I_1 + I_2) * H \;=\; (I_1 * H) + (I_2 * H). \tag{5.19}$$

It may be surprising, however, that simply *adding* a constant (scalar) value b to the image does *not* add to the convolved result by the same amount,

$$(b + I) * H \;\neq\; b + (I * H), \tag{5.20}$$

and is thus not part of the linearity property. While linearity is an important theoretical property, one should note that in practice "linear" filters are often only partially linear because of rounding errors or a limited range of output values.

Associativity

Linear convolution is associative, meaning that the order of successive filter operations is irrelevant:

$$A * (B * C) = (A * B) * C. \tag{5.21}$$

Thus multiple successive filters can be applied in any order, and multiple filters can be arbitrarily combined into new filters.

5.3.3 Separability of Linear Filters

If a convolution kernel H can be expressed as the convolution of multiple kernels itself,

$$H = H_1 * H_2 * \ldots * H_n,$$

then (as a consequence of Eqn. (5.21)) the filter operation $I * H$ may be performed as a sequence of convolutions with the constituting kernels,

$$\begin{aligned} I * H &= I * (H_1 * H_2 * \ldots * H_n) \\ &= \big(\ldots((I * H_1) * H_2) * \ldots * H_n\big). \end{aligned} \tag{5.22}$$

Depending upon the type of decomposition, this may result in significant computational savings.

x/y-separability

The possibility of separating a two-dimensional kernel H into a pair of one-dimensional kernels H_x, H_y is of particular relevance and is used in many practical applications. Let us assume, as a simple example, that the filter is composed of the one-dimensional kernels H_x and H_y with

$$H_x = \begin{bmatrix} 1 & 1 & \mathbf{1} & 1 & 1 \end{bmatrix} \quad \text{and} \quad H_y = \begin{bmatrix} 1 \\ \mathbf{1} \\ 1 \end{bmatrix}, \tag{5.23}$$

respectively.[5] If these filters are applied sequentially to the image I,

$$I' \leftarrow (I * H_x) * H_y = I * \underbrace{(H_x * H_y)}_{H_{xy}}, \tag{5.24}$$

then according to Eqn. (5.22) this is equivalent to applying the composite filter

$$H_{xy} = H_x * H_y = \begin{bmatrix} 1 & 1 & 1 & 1 & 1 \\ 1 & 1 & \mathbf{1} & 1 & 1 \\ 1 & 1 & 1 & 1 & 1 \end{bmatrix}. \tag{5.25}$$

[5] The bold values in the matrices mark the filters' coordinate origins (hot spots).

Thus this two-dimensional 5×3 box filter H_{xy} can be constructed from two one-dimensional filters of lengths 5 and 3, respectively (which is obviously true for box filters of any size). But what is the advantage of this? In the case above, the required amount of processing is $5 \cdot 3 = 15$ steps per image pixel for the 2D filter H_{xy} as compared with $5 + 3 = 8$ steps for the two separate 1D filters, a reduction of almost 50%. In general, the number of operations for a 2D filter grows *quadratically* with the filter size (side length) but only *linearly* if the filter is x/y-separable. Clearly, separability is an eminent bonus for the implementation of large linear filters (see also Sec. 5.5.1).

Separable Gaussian filters

In general, a two-dimensional filter is x/y-separable if (as in the example above) the filter function $H(i,j)$ can be expressed as the outer product ($\otimes$) of two one-dimensional functions,

$$H_{x,y}(i,j) = (H_x \otimes H_y)\,(i,j) = H_x(i) \cdot H_y(j), \tag{5.26}$$

because in this case the resulting function also corresponds to the convolution product $H_{x,y} = H_x * H_y$. A prominent example is the widely employed two-dimensional Gaussian function $G_\sigma(x,y)$ Eqn. (5.12), which can be expressed as the product

$$G_\sigma(x,y) \;=\; e^{-\frac{x^2+y^2}{2\sigma^2}} \;=\; e^{-\frac{x^2}{2\sigma^2}} \cdot e^{-\frac{y^2}{2\sigma^2}} \;=\; g_\sigma(x) \cdot g_\sigma(y). \tag{5.27}$$

Thus a two-dimensional Gaussian filter $H^{G,\sigma}$ can be implemented by a pair of one-dimensional Gaussian filters $H_x^{g,\sigma}$, $H_y^{g,\sigma}$ as

$$I' \leftarrow I * H^{G,\sigma} = I * H_x^{g,\sigma} * H_y^{g,\sigma}. \tag{5.28}$$

With different σ-values along the x and y axes, elliptical 2D Gaussians can be realized as separable filters in the same fashion.

The Gaussian function decays relatively slowly with increasing distance from the center. To avoid visible truncation errors, discrete approximations of the Gaussian should have a sufficiently large extent of about $\pm 2.5\,\sigma$ to $\pm 3.5\,\sigma$ samples. For example, a discrete 2D Gaussian with "radius" $\sigma = 10$ requires a minimum filter size of 51×51 pixels, in which case the x/y-separable version can be expected to run about 50 times faster than the full 2D filter. The Java method `makeGaussKernel1d()` in Prog. 5.4 shows how to dynamically create a one-dimensional Gaussian filter kernel with an extent of $\pm 3\,\sigma$ (i. e., a vector of odd length $6\,\sigma + 1$). As an example, this method is used for implementing "unsharp masking" filters where relatively large Gaussian kernels may be required (see Prog. 6.1 in Sec. 6.6.2).

```
float[] makeGaussKernel1d(double sigma) {

  // create the kernel
  int center = (int) (3.0*sigma);
  float[] kernel = new float[2*center+1]; // odd size

  // fill the kernel
  double sigma2 = sigma * sigma;          // σ^2
  for (int i=0; i<kernel.length; i++) {
    double r = center - i;
    kernel[i] = (float) Math.exp(-0.5 * (r*r) / sigma2);
  }

  return kernel;
}
```

Program 5.4 Dynamic creation of one-dimensional Gaussian filter kernels. For a given σ, the Java method `makeGaussKernel1d()` returns a discrete 1D Gaussian filter kernel (`float` array) large enough to avoid truncation effects.

5.3.4 Impulse Response of a Filter

Linear convolution is a binary operation involving two functions as its operands; it also has a "neutral element", which of course is a function, too. The *impulse* or *Dirac* function $\delta()$ is neutral under convolution; i. e.,

$$I * \delta = I. \tag{5.29}$$

In the discrete, two-dimensional case, the impulse function is defined as

$$\delta(u,v) = \begin{cases} 1 & \text{for } u = v = 0 \\ 0 & \text{otherwise.} \end{cases} \tag{5.30}$$

Interpreted as an image, this function is merely a single bright pixel (with value 1) at the coordinate origin contained in a dark (zero value) plane of infinite extent (Fig. 5.10).

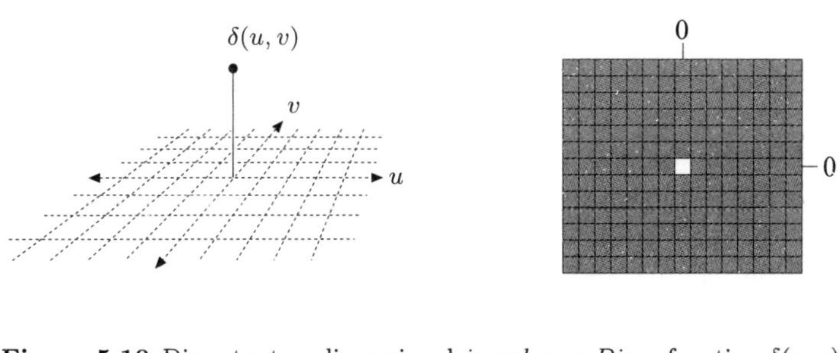

Figure 5.10 Discrete, two-dimensional *impulse* or *Dirac* function $\delta(u,v)$.

When the Dirac function is used as the filter kernel in a linear convolution as in Eqn. (5.29), the result is identical to the original image (Fig. 5.11). The reverse situation is more interesting, however, where some filter H is applied to the impulse δ as the input function. What happens? Since convolution is commutative (Eqn. (5.17)) it is evident that

$$H * \delta = \delta * H = H \tag{5.31}$$

and thus the result of this filter operation is identical to the filter H itself (Fig. 5.12)! While sending an impulse into a linear filter to obtain its filter function may seem paradoxical at first, it makes sense if the properties (coefficients) of the filter H are unknown. Assuming that the filter is actually linear, complete information about this filter is obtained by injecting only a single impulse and measuring the result, which is called the "impulse response" of the filter. Among other applications, this technique is used for measuring the behavior of optical systems (e. g., lenses), where a point light source serves as the impulse and the result—a distribution of light energy—is called the "point spread function" (PSF) of the system.

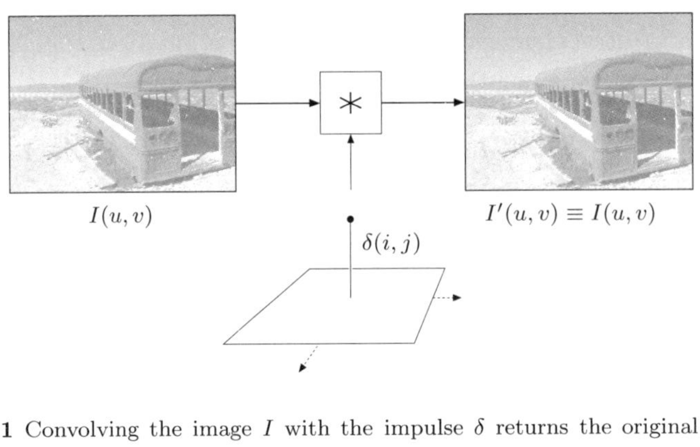

Figure 5.11 Convolving the image I with the impulse δ returns the original unmodified image.

5.4 Nonlinear Filters

Linear filters have an important disadvantage when used for smoothing or removing noise: all image structures, including points, edges, and lines, are also blurred, and the quality of the whole image is evenly reduced (Fig. 5.13). This effect cannot be avoided, and thus the use of linear filters for these kinds of tasks (noise removal in particular) is limited. In the following, we investigate certain nonlinear filters to see if they can offer any better solution to this problem.

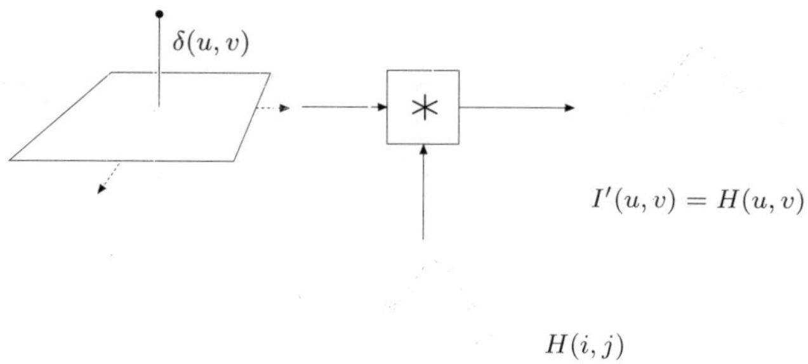

Figure 5.12 The linear filter H with the impulse δ as the input yields the filter H as the result.

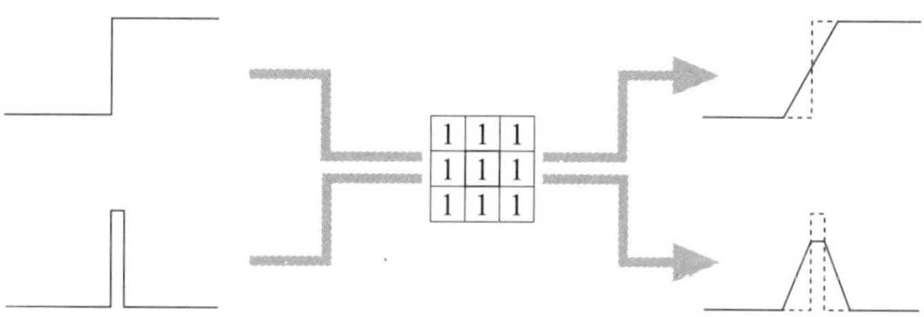

Figure 5.13 Any image structure is blurred by a linear smoothing filter. Important image structures such as step edges (top) or thin lines (bottom) are widened, and the local contrast is reduced.

5.4.1 Minimum and Maximum Filters

Like all other filters, nonlinear filters compute the result at some image position (u, v) from the pixels inside the moving region $R_{u,v}$ of the original image. The filters are called "nonlinear" because the source pixel values are combined by some nonlinear function. The simplest of all nonlinear filters are the *minimum* and *maximum* filters, defined as

$$I'(u,v) \leftarrow \min\{I(u+i,v+j) \mid (i,j) \in R\}, \quad (5.32)$$

$$I'(u,v) \leftarrow \max\{I(u+i,v+j) \mid (i,j) \in R\}, \quad (5.33)$$

where R denotes the filter region (set of filter coordinates), usually a square of size 3×3 pixels. Figure 5.14 illustrates the effects of a one-dimensional minimum filter on various local signal structures.

Figure 5.15 shows the results of applying 3×3 pixel minimum and maximum filters to a grayscale image corrupted with "salt-and-pepper" noise (i. e.,

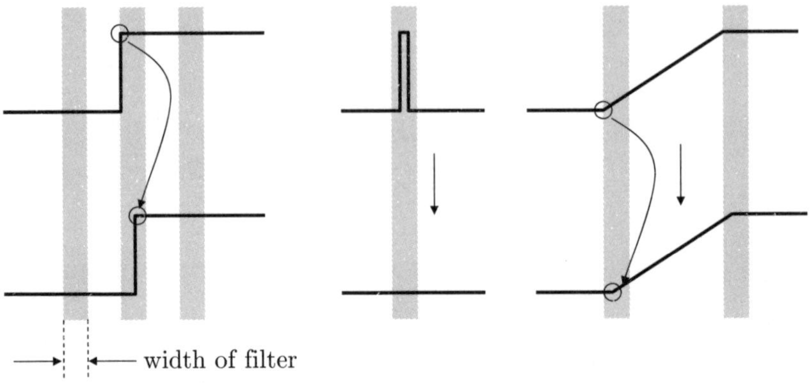

Figure 5.14 Effects of a one-dimensional minimum filter on various local signal structures. Original signal (top) and result after filtering (bottom), where the colored bars indicate the extent of the filter. The step edge (a) and the linear ramp (c) are shifted to the right by half the filter width, and the narrow pulse (b) is completely removed.

randomly placed white and black dots), respectively. Obviously the minimum filter removes the white (salt) dots because any single white pixel within the 3×3 filter region is replaced by one of its surrounding pixels with a smaller value. Notice, however, that the minimum filter at the same time widens all the dark structures in the image.

The reverse effects can be expected from the *maximum* filter. Any single bright pixel is a local maximum as soon as it is contained in the filter region R. White dots (and all other bright image structures) are thus widened to the size of the filter, while now the dark ("pepper") dots disappear.

5.4.2 Median Filter

It is impossible of course to design a filter that removes any noise but keeps all the important image structures intact because no filter can discriminate which image content is important to the viewer and which is not. The popular median filter is at least a good step in this direction.

The median filter replaces every image pixel by the *median* of the pixels in the corresponding filter region R,

$$I'(u,v) \leftarrow \text{median}\,\{I(u+i, v+j) \mid (i,j) \in R\}\,. \tag{5.34}$$

The median of a sequence of $2K+1$ values p_i is defined as the center value p_K

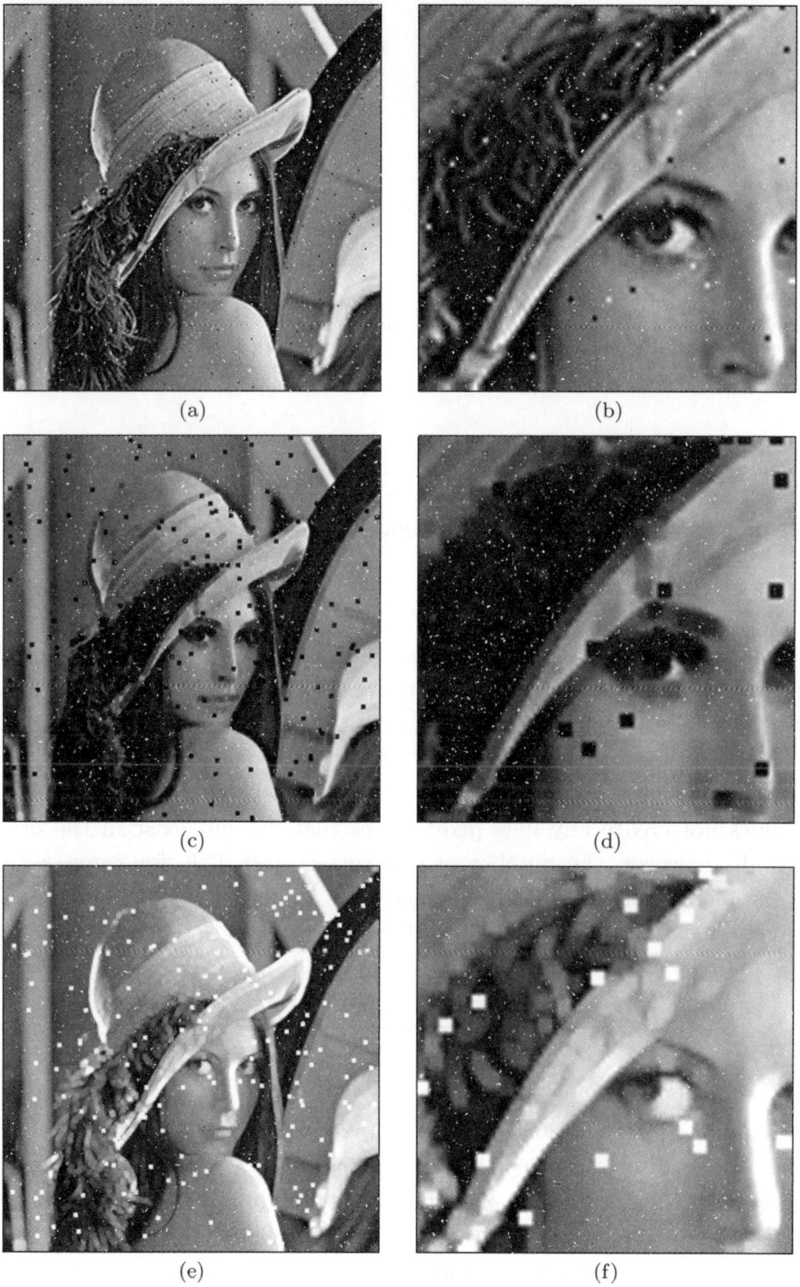

Figure 5.15 Minimum and maximum filters applied to a grayscale image. The original image (a, b) is corrupted with "salt-and-pepper" noise. The 3×3 pixel *minimum* filter eliminates the bright dots and widens all dark image structures (c, d). The *maximum* filter shows the exact opposite effects (e, f).

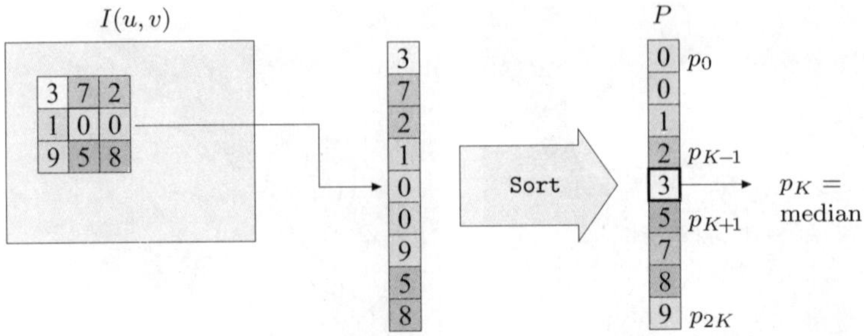

Figure 5.16 Computation of a 3×3 pixel median filter. The nine pixel values extracted from the 3×3 image region are arranged as a vector (P) that is sorted, and the resulting center value is taken as the median.

after the sequence $P = (p_0, \ldots, p_{2K})$ is *sorted*; i. e.,

$$\operatorname{median}\big(\underbrace{p_0, p_1, \ldots, p_{K-1}}_{\substack{K \text{values} \\ p_i \leq p_K}}, p_K, \underbrace{p_{K+1}, \ldots, p_{2K}}_{\substack{K \text{values} \\ p_i \geq p_K}}\big) = p_K, \tag{5.35}$$

where $p_i \leq p_{i+1}$ (for $0 \leq i < 2K$). Figure 5.16 demonstrates the computation of the median filter or a filter region of size 3×3 pixels.

Equation (5.35) defines the median of an *odd*-sized set of values, and if the side length of the rectangular filters is odd (which is usually the case), then the number of elements in the filter region is odd as well. In this case, the median filter does not create any new pixel values that did not exist in the original image. If, however, the number of elements is *even* ($2K$ for some $K > 0$), then the median of the sorted sequence $P = (p_0, \ldots, p_{2K-1})$ is defined as the arithmetic mean of the two middle values p_{K-1} and p_K,

$$\operatorname{median}\big(\underbrace{p_0, \ldots, p_{K-1}}_{\substack{K \text{values} \\ p_i \leq p_K}}, \underbrace{p_K, \ldots, p_{2K-1}}_{\substack{K \text{values} \\ p_i \geq p_K}}\big) = \tfrac{1}{2} \cdot (p_{K-1} + p_K). \tag{5.36}$$

Because of the interpolation above, new pixel values are generally introduced by the median filter if the region is of even size.

Figure 5.17 illustrates the effects of a 3×3 pixel median filter on selected two-dimensional image structures. In particular, very small structures (smaller than half the filter size) are eliminated, but all other structures remain largely unchanged. Finally, Fig. 5.18 compares the results of median filtering with a linear-smoothing filter. A sample Java implementation of the median filter, whose principal structure is identical to the 3×3 pixel linear filter in Prog. 5.2, is shown in Prog. 5.5.

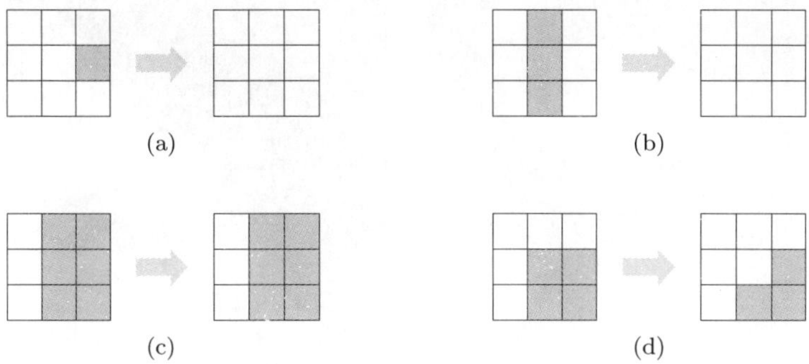

Figure 5.17 Effects of a 3×3 pixel median filter on two-dimensional image structures. Isolated dots are eliminated (a), as are thin lines (b). The step edge remains unchanged (c), while a corner is rounded off (d).

5.4.3 Weighted Median Filter

The median is a rank order statistic, and in a sense the "majority" of the pixel values involved determine the result. A single exceptionally high or low value (an "outlier") cannot influence the result much but only shift the result up or down to the next value. Thus the median (in contrast to the linear average) is considered a "robust" measure. In an ordinary median filter, each pixel in the filter region has the same influence, regardless of its distance from the center.

The weighted median filter assigns individual weights to the positions in the filter region, which can be interpreted as the "number of votes" for the corresponding pixel values. Similar to the coefficient matrix H of a linear filter, the distribution of weights is specified by a *weight matrix* W, with $W(i,j) \in \mathbb{N}$. To compute the result of the modified filter, each pixel value $I(u+i, v+j)$ involved is inserted $W(i,j)$ times into the extended pixel vector

$$Q = (p_0, \ldots, p_{L-1}) \quad \text{of length} \quad L = \sum_{(i,j) \in R} W(i,j).$$

This vector is then sorted, and the resulting center value is taken as the median, as in the standard median filter. Figure 5.19 illustrates the computation of the weighted median filter using the 3×3 weight matrix

$$W(i,j) = \begin{bmatrix} 1 & 2 & 1 \\ 2 & \mathbf{3} & 2 \\ 1 & 2 & 1 \end{bmatrix}, \tag{5.37}$$

which requires an extended pixel vector of length $L = 15$, equal to the sum of the weights in W.

Of course this method may also be used to implement ordinary median filters of nonrectangular shape; for example, a *cross-shaped* median filter with

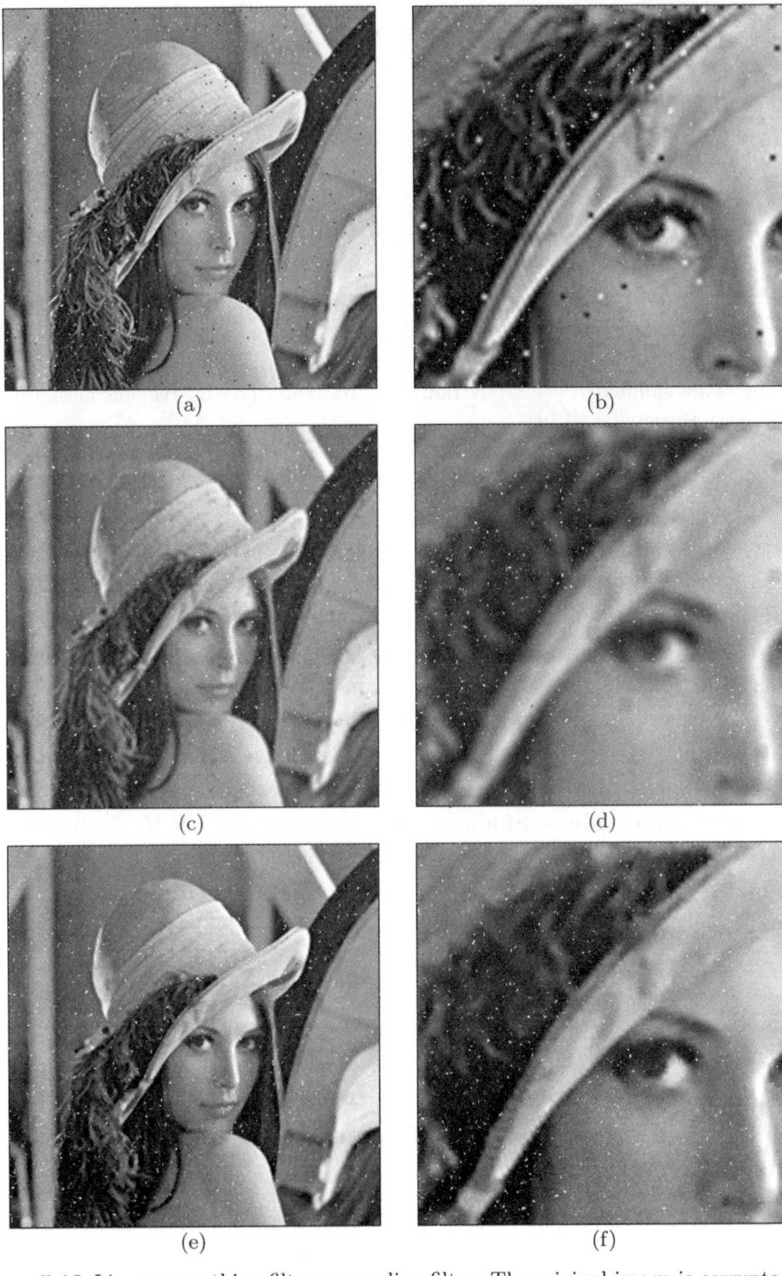

Figure 5.18 Linear smoothing filter vs. median filter. The original image is corrupted with "salt-and-pepper" noise (a, b). The linear 3×3 pixel box filter (c, d) reduces the bright and dark peaks to some extent but is unable to remove them completely. In addition, the entire image is blurred. The median filter (e, f) effectively eliminates the noise dots and also keeps the remaining structures largely intact. However, it also creates small spots of flat intensity that noticeably affect the sharpness.

```
1 import ij.*;
2 import ij.plugin.filter.PlugInFilter;
3 import ij.process.*;
4 import java.util.Arrays;
5
6 public class Filter_Median3x3 implements PlugInFilter {
7     final int K = 4; // filter size
8
9     public void run(ImageProcessor orig) {
10        int w = orig.getWidth();
11        int h = orig.getHeight();
12        ImageProcessor copy = orig.duplicate();
13
14        // vector to hold pixels from 3x3 neighborhood
15        int[] P = new int[2*K+1];
16
17        for (int v = 1; v <= h-2; v++) {
18            for (int u = 1; u <= w-2; u++) {
19                // fill the pixel vector P for filter position u, v
20                int k = 0;
21                for (int j = -1; j <= 1; j++) {
22                    for (int i = -1; i <= 1; i++) {
23                        P[k] = copy.getPixel(u+i, v+j);
24                        k++;
25                    }
26                }
27                // sort pixel vector and take the center element
28                Arrays.sort(P);
29                orig.putPixel(u, v, P[K]);
30            }
31        }
32    }
33
34 } // end of class Filter_Median3x3
```

Program 5.5 A 3×3 median filter (ImageJ plugin). An array `P` of type `int` is defined (line 15) to hold the 9 pixels for each filter position (u, v). This array is sorted by using the Java utility method `Arrays.sort()` in line 28. The center element of the sorted vector (`P[K]`) is taken as the median value and stored in the original image (line 29).

the weight matrix

$$W^{+}(i,j) = \begin{bmatrix} 0 & 1 & 0 \\ 1 & \mathbf{1} & 1 \\ 0 & 1 & 0 \end{bmatrix}. \tag{5.38}$$

Not every arrangement of weights is useful, however. In particular, if the weight assigned to the center pixel is greater than the sum of all other weights, then that pixel would always have the "majority" and dictate the resulting value, thus inhibiting any filter effect.

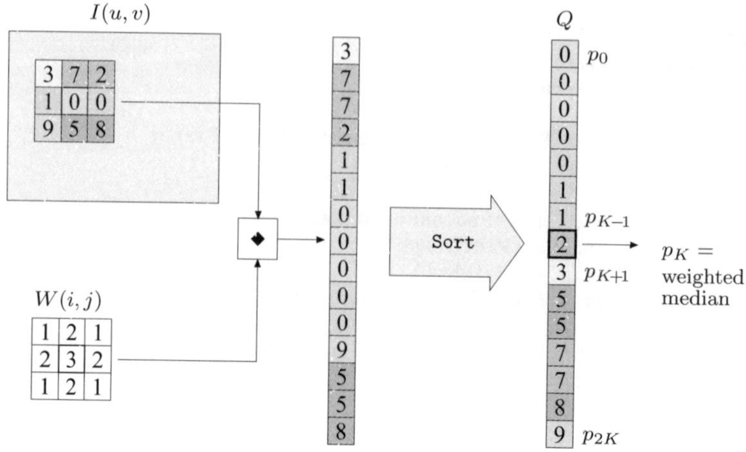

Figure 5.19 Weighted median example. Each pixel value is inserted into the extended pixel vector multiple times, as specified by the weight matrix W. For example, the value 0 from the center pixel is inserted three times (since $W(0,0) = 3$) and the pixel value 7 twice. The pixel vector is sorted and the center value (2) is taken as the median.

5.4.4 Other Nonlinear Filters

Median and weighted median filters are two examples of nonlinear filters that are easy to describe and frequently used. Since "nonlinear" refers to anything that is not linear, there are a multitude of filters that fall into this category, including the morphological filters for binary and grayscale images, which are discussed in Ch. 7. Other types of nonlinear filters, such as the corner detector described in Vol. 2 [6, Ch. 4], are often described algorithmically and thus defy a simple, compact description.

In contrast to the linear case, there is usually no "strong theory" for nonlinear filters that could, for example, describe the relationship between the sum of two images and the results of a median filter, as does Eqn. (5.19) for linear convolution. Similarly, not much (if anything) can be stated in general about the effects of nonlinear filters in frequency space.

5.5 Implementing Filters

5.5.1 Efficiency of Filter Programs

Computing the results of filters is computationally expensive in most cases, especially with large images, large filter kernels, or both. Given an image of size $M \times N$ and a filter kernel of size $(2K+1) \times (2L+1)$, a direct implementation requires

$$2K \cdot 2L \cdot M \cdot N = 4KLMN$$

operations, namely multiplications and additions (in the case of a linear filter). Thus, if both the image and the filter are simply assumed to be of size $N \times N$, the time complexity[6] of direct filtering is $\mathcal{O}(N^4)$. As described in Sec. 5.3.3, substantial savings are possible when large, two-dimensional filters can be decomposed (separated) into smaller, possibly one-dimensional filters.

The programming examples in this chapter are deliberately designed to be simple and easy to understand, and none of the solutions shown are particularly efficient. Possibilities for tuning and code optimization exist in many places. It is particularly important to move all unnecessary instructions out of inner loops if possible because these are executed most often. This applies especially to "expensive" instructions, such as method invocations, which may be relatively time-consuming (particularly in Java).

In the examples, we have intentionally used the ImageJ standard methods `getPixel()` for reading and `putPixel()` for writing image pixels, which is the simplest and safest approach to access image data but also the slowest, of course. Substantial speed can be gained by using the quicker read and write methods `get()` and `set()` defined for class `ImageProcessor` and its subclasses. Note, however, that these methods do not check if the passed image coordinates are valid. Maximum performance can be obtained by accessing the pixel arrays directly.[7]

5.5.2 Handling Image Borders

As mentioned briefly in Sec. 5.2.2, the image borders require special attention in most filter implementations. We have argued that theoretically no filter results can be computed at positions where the filter matrix is not fully contained in the image array. Thus any filter operation would reduce the size of the resulting image, which is not acceptable in most applications. While no formally correct remedy exists, there are several more or less practical methods for handling the remaining border regions:

Method 1: Set the unprocessed pixels at the borders to some constant value (e. g., "black"). This is certainly the simplest method, but not acceptable in many situations because the image size is incrementally reduced by every filter operation.

Method 2: Set the unprocessed pixels to the original (unfiltered) image values. Usually the results are unacceptable, too, due to the noticeable difference between filtered and unprocessed image parts.

---

[6] See Appendix A (p. 235) for a short description of the $\mathcal{O}()$ notation.

[7] See the ImageJ Short Reference [5, Chap. 7] for details.

Method 3: Extend the image by "padding" additional pixels around it (Fig. 5.20) and filter the border regions, too, assuming that:

A. The pixels outside the image have a *constant value* (e. g., "black" or "gray"; Fig. 5.20 (a)). This may produce strong artifacts at the image borders, particularly when large filters are used.

B. The *border pixels extend* beyond the image boundaries (Fig. 5.20 (b)). Only minor artifacts can be expected at the borders. The method is also simple to compute and is thus often considered the method of choice.

C. The *image is mirrored* at each of its four boundaries (Fig. 5.20 (c)). The results will be similar to those of the previous method unless very large filters are used.

D. The *image repeats periodically* in the horizontal and vertical directions (Fig. 5.20 (d)). This may seem strange at first, and also the results are generally not satisfactory. However, in discrete spectral analysis, the image is implicitly treated as a periodic function, too.[8] Thus, if the image is filtered in the frequency domain, the results will be equal to filtering in the space domain under this repetitive model.

None of these methods is perfect and, as usual, the right choice depends upon the type of image and the filter applied. Notice also that the special treatment of the image borders may sometimes require more programming effort (and computing time) than the processing of the interior image.

5.5.3 Debugging Filter Programs

Experience shows that programming errors can hardly ever be avoided, even by experienced practitioners. Unless errors occur during execution (usually caused by trying to access nonexistent array elements), filter programs always "do something" to the image that may be similar but not identical to the expected result. To assure that the code operates correctly, it is not advisable to start with full, large images but first to experiment with small test cases for which the outcome can easily be predicted. Particularly when implementing linear filters, a first "litmus test" should always be to inspect the impulse response of the filter (as described in Sec. 5.3.4) before processing any real images.

5.6 Filter Operations in ImageJ

ImageJ offers a collection of readily available filter operations, many of them contributed by other authors using different styles of implementation. Most of

[8] This comment refers to topics covered in Vol. 2 [6, Ch. 7].

(a) (b) (c) (d)

Figure 5.20 Methods for padding the image to facilitate filtering along the borders. The assumption is that the (nonexisting) pixels outside the original image are either set to some constant value (a), take on the value of the closest border pixel (b), are mirrored at the image boundaries (c), or repeat periodically along the coordinate axes (d).

the available operations can also be invoked via ImageJ's Process menu.

5.6.1 Linear Filters

Filters based on linear convolution are implemented by the ImageJ plugin class `ij.plugin.filter.Convolver`, which offers useful "public" methods in addition to the standard `run()` method. Usage of this class is illustrated by the following example that convolves an 8-bit grayscale image with the filter kernel from Eqn. (5.7):

$$H(i,j) = \begin{bmatrix} 0.075 & 0.125 & 0.075 \\ 0.125 & \mathbf{0.2} & 0.125 \\ 0.075 & 0.125 & 0.075 \end{bmatrix}.$$

In our `run()` method below, we first define the filter matrix `H` as a *one-dimensional* `float` array (notice the syntax for the `float` constants "`0.075f`", etc.) and then create a new instance (`cv`) of class `Convolver` in line 8:

```
1  import ij.plugin.filter.Convolver;
2  ...
3  public void run(ImageProcessor I) {
4    float[] H = {              // filter array is one-dimensional!
5        0.075f, 0.125f, 0.075f,
6        0.125f, 0.200f, 0.125f,
7        0.075f, 0.125f, 0.075f };
8    Convolver cv = new Convolver();
9    cv.setNormalize(false);  // do not use filter normalization
10   cv.convolve(I, H, 3, 3); // apply the filter H to I
11 }
```

The invocation of the method `convolve()` in line 10 applies the filter `H` to the image `I`. It requires two additional arguments for the dimensions of the filter matrix since `H` is passed as a one-dimensional array. The image `I` is destructively modified by the `convolve` operation.

In this case, one could have also used the nonnormalized, integer-valued filter matrix given in Eqn. (5.10) because `convolve()` normalizes the given filter automatically (after `cv.setNormalize(true)`).

5.6.2 Gaussian Filters

The ImageJ class `ij.plugin.filter.GaussianBlur` implements a simple Gaussian blur filter with arbitrary radius (σ). The filter uses separable one-dimensional Gaussians as described in Sec. 5.3.3. Here is an example showing its application with the radius $\sigma = 2.5$:

```
import ij.plugin.filter.GaussianBlur;
...
public void run(ImageProcessor ip) {
  GaussianBlur gb = new GaussianBlur();
  double radius = 2.5;
  gb.blur(ip, radius);
}
```

An alternative implementation of separable Gaussian filters can be found in Prog. 6.1 (see p. 154), which uses the method `makeGaussKernel1d()` defined in Prog. 5.4 (page 115) for dynamically computing the required 1D filter kernels.

5.6.3 Nonlinear Filters

A small set of nonlinear filters is implemented in the ImageJ class `ij.plugin.filter.RankFilters`, including the minimum, maximum, and standard median filters. The filter region is (approximately) circular with variable radius.

Here is an example that applies three different filters with the same radius in sequence:

```
import ij.plugin.filter.RankFilters;
...
public void run(ImageProcessor ip) {
  RankFilters rf = new RankFilters();
  double radius = 3.5;
  rf.rank(ip, radius, RankFilters.MIN);   // minimum filter
  rf.rank(ip, radius, RankFilters.MAX);   // maximum filter
  rf.rank(ip, radius, RankFilters.MEDIAN);   // median filter
}
```

5.7 Exercises

Exercise 5.1
Explain why the "custom filter" in Adobe Photoshop (Fig. 5.6) is not strictly a linear filter.

Exercise 5.2
Determine the possible maximum and minimum results (pixel values) for a linear filter with

$$H(i,j) = \begin{bmatrix} -1 & -2 & 0 \\ -2 & \mathbf{0} & 2 \\ 0 & 2 & 1 \end{bmatrix}$$

when applied to an 8-bit grayscale image (with pixel values in the range $[0, 255]$). Assume that no clamping of the results occurs.

Exercise 5.3
Modify the ImageJ plugin shown in Prog. 5.3 such that the image borders are processed as well. Use one of the methods for extending the image outside its boundaries as described in Sec. 5.5.2.

Exercise 5.4
Show that a standard box filter is not isotropic (i. e., does not smooth the image identically in all directions).

Exercise 5.5
Explain why the clamping of results to a limited range of pixel values may violate the linearity property (Sec. 5.3.2) of linear filters.

Exercise 5.6
Compare the number of processing steps required for non-separable linear filters and x/y-separable filters sized 5×5, 11×11, 25×25, and 51×51 pixels. Compute the speed gain resulting from separability in each case.

Exercise 5.7

Implement a weighted median filter (Sec. 5.4.3) as an ImageJ plugin, specifying the weights as a constant, two-dimensional `int` array. Test the filter on suitable images and compare the results with those from a standard median filter. Explain why, for example, the weight matrix

$$W(i,j) = \begin{bmatrix} 0 & 1 & 0 \\ 1 & \mathbf{5} & 1 \\ 0 & 1 & 0 \end{bmatrix}$$

does *not* make sense.

Exercise 5.8

Verify the properties of the *impulse* function with respect to linear filters Eqn. (5.31). Create a black image with a white pixel at its center and use this image as the two-dimensional impulse. See if linear filters really deliver the filter matrix H as their impulse response.

Exercise 5.9

Describe the effect of a linear filter with the following filter matrix:

$$H(i,j) = \begin{bmatrix} 0 & 0 & 0 \\ 0 & \mathbf{0} & 1 \\ 0 & 0 & 0 \end{bmatrix}.$$

Exercise 5.10

Design a linear filter (matrix) that creates a horizontal blur over a length of 7 pixels, thus simulating the effect of camera movement during exposure.

Exercise 5.11

Program your own ImageJ plugin that implements a Gaussian smoothing filter with variable filter width (radius σ). The plugin should dynamically create the required filter kernels with a size of at least 5σ in both directions. Make use of the fact that the Gaussian function is x/y-separable (see Sec. 5.3.3).

Exercise 5.12

The "*Laplacian of Gaussian*" (LoG) filter (Fig. 5.8) is based on the sum of the second derivatives of the two-dimensional Gaussian. It is defined as

$$\mathrm{LoG}_\sigma(x,y) = -\left(\frac{x^2 + y^2 - 2\sigma^2}{\sigma^4}\right) \cdot e^{-\frac{x^2+y^2}{2\sigma^2}}.$$

Implement the LoG filter as an ImageJ plugin of variable width (σ), analogous to Exercise 5.11. Find out if the LoG function is x/y-separable.

6
Edges and Contours

Prominent image "events" originating from local changes in intensity or color, such as edges and contours, are of high importance for the visual perception and interpretation of images. The perceived amount of information in an image appears to be directly related to the distinctiveness of the contained structures and discontinuities. In fact, edge-like structures and contours seem to be so important for our human visual system that a few lines in a caricature or illustration are often sufficient to unambiguously describe an object or a scene. It is thus no surprise that the enhancement and detection of edges has been a traditional and important topic in image processing as well. In this chapter, we first look at simple methods for localizing edges and then attend to the related issue of image sharpening.

6.1 What Makes an Edge?

Edges and contours play a dominant role in human vision and probably in many other biological vision systems as well. Not only are edges visually striking, but it is often possible to describe or reconstruct a complete figure from a few key lines, as the example in Fig. 6.1 shows. But how do edges arise, and how can they be technically localized in an image?

Edges can roughly be described as image positions where the local intensity changes distinctly along a particular orientation. The stronger the local intensity change, the higher is the evidence for an edge at that position. In mathematics, the amount of change with respect to spatial distance is known

W. Burger, M.J. Burge, *Principles of Digital Image Processing*, Undergraduate Topics in Computer Science, DOI 10.1007/978-1-84800-191-6_6, © Springer-Verlag London Limited, 2009

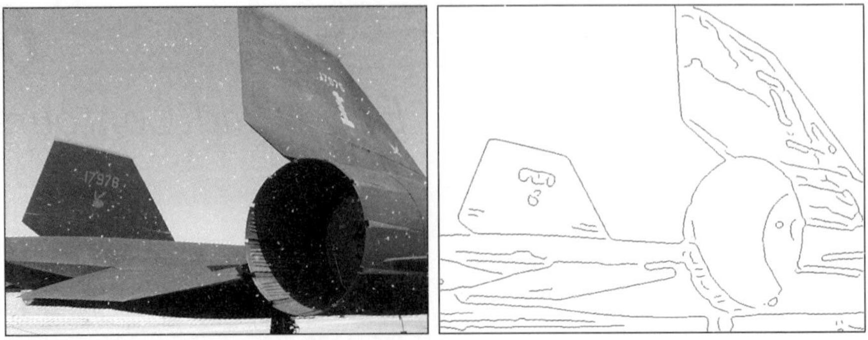

(a) (b)

Figure 6.1 Edges play an important role in human vision. Original image (a) and edge image (b).

as the first derivative of a function, and we thus start with this concept to develop our first simple edge detector.

6.2 Gradient-Based Edge Detection

For simplicity, we first investigate the situation in only one dimension, assuming that the image contains a single bright region at the center surrounded by a dark background (Fig. 6.2 (a)). In this case, the intensity profile along one image line would look like the one-dimensional function $f(x)$, as shown in Fig. 6.2 (b). Taking the first derivative of the function f,

$$f'(x) = \frac{df}{dx}(x) \tag{6.1}$$

results in a positive swing at those positions where the intensity rises and a negative swing where the value of the function drops (Fig. 6.2 (c)).

Unlike in the continuous case, however, the first derivative is undefined for a *discrete* function $f(u)$ (such as the line profile of a real image), and some method is needed to estimate it. Figure 6.3 gives the basic idea, again for the one-dimensional case: the first derivative of a continuous function at position x can be interpreted as the slope of its *tangent* at this position. One simple method for roughly approximating the slope of the tangent for a *discrete* function $f(u)$ at position u is to fit a straight line through the neighboring function values $f(u-1)$ and $f(u+1)$,

$$\frac{df}{du}(u) \approx \frac{f(u+1) - f(u-1)}{(u+1) - (u-1)} = \frac{f(u+1) - f(u-1)}{2} . \tag{6.2}$$

The same method can be applied of course in the vertical direction to estimate the first derivative along the y-axis; i. e., along the image columns.

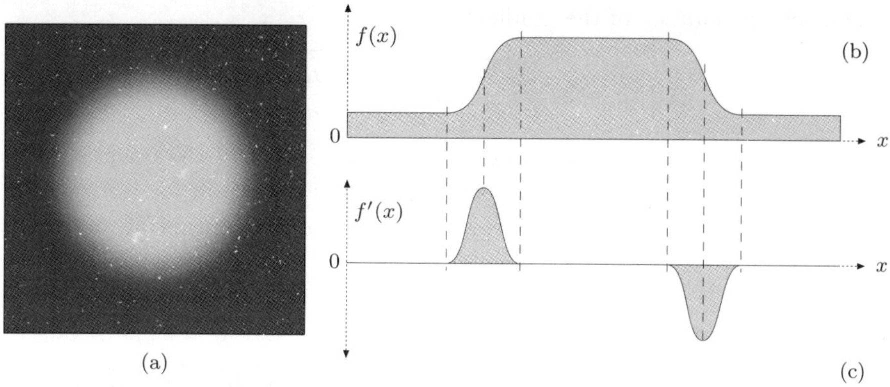

Figure 6.2 Sample image and first derivative in one dimension: original image (a), horizontal intensity profile $f(x)$ along the center image line (b), and first derivative $f'(x)$ (c).

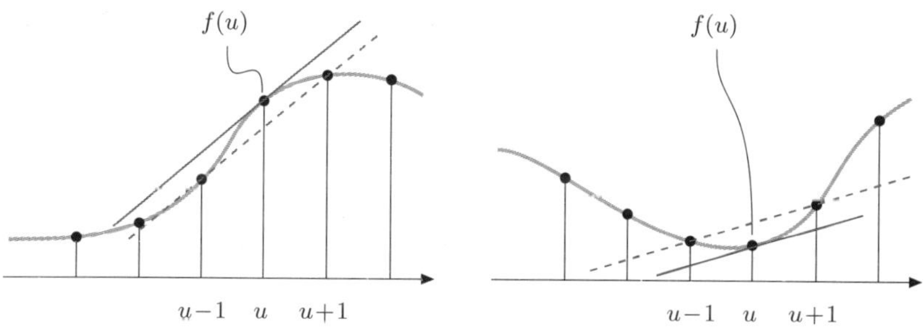

Figure 6.3 Estimating the first derivative of a discrete function. The slope of the straight (dashed) line between the neighboring function values $f(u-1)$ and $f(u+1)$ is taken as the estimate for the slope of the tangent (i. e., the first derivative) at $f(u)$.

6.2.1 Partial Derivatives and the Gradient

A derivative of a multidimensional function taken along one of its coordinate axes is called a *partial derivative*; for example,

$$\frac{\partial I}{\partial u}(u,v) \qquad \text{and} \qquad \frac{\partial I}{\partial v}(u,v) \tag{6.3}$$

are the partial derivatives of the image function $I(u, v)$ along the u and v axes, respectively.[1] The function

$$\nabla I(u,v) = \begin{bmatrix} \frac{\partial I}{\partial u}(u,v) \\ \frac{\partial I}{\partial v}(u,v) \end{bmatrix} \tag{6.4}$$

[1] ∂ denotes the *partial derivative* or "del" operator.

is called the *gradient vector* (or "gradient" for short) of the function I at position (u, v). The *magnitude* of the gradient,

$$|\nabla I|(u,v) = \sqrt{\left(\frac{\partial I}{\partial u}(u,v)\right)^2 + \left(\frac{\partial I}{\partial v}(u,v)\right)^2}, \tag{6.5}$$

is invariant under image rotation and thus independent of the orientation of the underlying image structures. This property is important for isotropic localization of edges, and thus $|\nabla I|$ is the basis of many practical edge detection methods.

6.2.2 Derivative Filters

The components of the gradient function (Eqn. (6.4)) are simply the first derivatives of the image lines (Eqn. (6.1)) and columns along the horizontal and vertical axes, respectively. The approximation of the first horizontal derivatives (Eqn. (6.2)) can be easily implemented by a linear filter (see Sec. 5.2) with the coefficient matrix

$$H_x^D = \begin{bmatrix} -0.5 & \mathbf{0} & 0.5 \end{bmatrix} = 0.5 \cdot \begin{bmatrix} -1 & \mathbf{0} & 1 \end{bmatrix}, \tag{6.6}$$

where the coefficients -0.5 and $+0.5$ apply to the image elements $I(u-1, v)$ and $I(u+1, v)$, respectively. Notice that the center pixel $I(u, v)$ itself is weighted with the zero coefficient and is thus ignored. Similarly, the vertical component of the gradient can be computed with the linear filter

$$H_y^D = \begin{bmatrix} -0.5 \\ \mathbf{0} \\ 0.5 \end{bmatrix} = 0.5 \cdot \begin{bmatrix} -1 \\ \mathbf{0} \\ 1 \end{bmatrix}. \tag{6.7}$$

Figure 6.4 shows the results of applying the gradient filters defined in Eqn. (6.6) and Eqn. (6.7) to a synthetic test image. The orientation dependence of the filter responses can be seen clearly. The horizontal gradient filter H_x^D reacts most strongly to rapid changes along the horizontal direction, (i. e., to *vertical* edges); analogously the vertical gradient filter H_y^D reacts most strongly to *horizontal* edges. The filter response is zero in flat image regions (shown gray in Fig. 6.4 (b, c)).

6.3 Edge Operators

The local gradient of the image function is the basis of many classical edge-detection operators. Practically, they only differ in the type of filter used for estimating the gradient components and the way these components are combined. In many situations, one is not only interested in the *strength* of edge

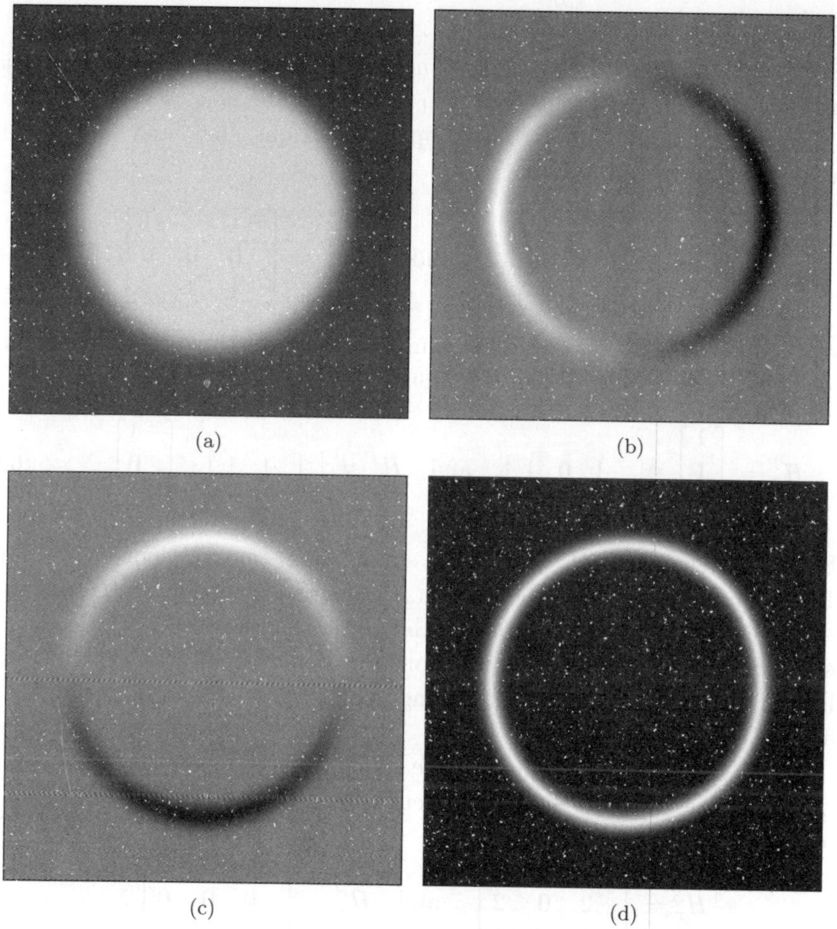

Figure 6.4 Partial derivatives of a two-dimensional function: synthetic image function I (a); approximate first derivatives in the horizontal direction $\partial I/\partial u$ (b) and the vertical direction $\partial I/\partial v$ (c); magnitude of the resulting gradient $|\nabla I|$ (d). In (b) and (c), the lowest (negative) values are shown black, the maximum (positive) values are white, and zero values are gray.

points but also in the local *direction* of the edge. Both types of information are contained in the gradient function and can be easily computed from the directional components. The following small collection describes some frequently used, simple edge operators that have been around for many years and are thus interesting from a historical perspective as well.

6.3.1 Prewitt and Sobel Operators

The edge operators by Prewitt and Sobel [10] are two classic methods that differ only marginally in the filters they use.

Gradient filters

The Prewitt and Sobel operators use linear filters that extend over three adjacent lines and columns, respectively, to counteract the noise sensitivity of the simple (single line/column) gradient operators (Eqns. (6.6) and (6.7)). The Prewitt operator uses the filters

$$H_x^P = \begin{bmatrix} -1 & 0 & 1 \\ -1 & \mathbf{0} & 1 \\ -1 & 0 & 1 \end{bmatrix} \quad \text{and} \quad H_y^P = \begin{bmatrix} -1 & -1 & -1 \\ 0 & \mathbf{0} & 0 \\ 1 & 1 & 1 \end{bmatrix}, \tag{6.8}$$

which compute the average gradient components across three neighboring lines or columns, respectively. When the filters are written in separated form,

$$H_x^P = \begin{bmatrix} 1 \\ \mathbf{1} \\ 1 \end{bmatrix} * \begin{bmatrix} -1 & \mathbf{0} & 1 \end{bmatrix} \quad \text{and} \quad H_y^P = \begin{bmatrix} 1 & \mathbf{1} & 1 \end{bmatrix} * \begin{bmatrix} -1 \\ \mathbf{0} \\ 1 \end{bmatrix}, \tag{6.9}$$

it becomes obvious that H_x^P performs a simple (box) smoothing over three lines before computing the x gradient (Eqn. (6.6)), and analogously H_y^P smooths over three columns before computing the y gradient (Eqn. (6.7)).[2] Because of the commutativity of linear convolution, this could equally be described the other way around, with smoothing being applied *after* the computation of the gradients.

The filters for the Sobel operator are almost identical; however, the smoothing part assigns higher weight to the current center line and column, respectively:

$$H_x^S = \begin{bmatrix} -1 & 0 & 1 \\ -2 & \mathbf{0} & 2 \\ -1 & 0 & 1 \end{bmatrix} \quad \text{and} \quad H_y^S = \begin{bmatrix} -1 & -2 & -1 \\ 0 & \mathbf{0} & 0 \\ 1 & 2 & 1 \end{bmatrix}. \tag{6.10}$$

The estimates for the local gradient components are obtained from the filter results by appropriate scaling:

$$\nabla I(u,v) \approx \frac{1}{6} \cdot \begin{bmatrix} (I * H_x^P)(u,v) \\ (I * H_y^P)(u,v) \end{bmatrix} \tag{6.11}$$

for the *Prewitt* operator and

$$\nabla I(u,v) \approx \frac{1}{8} \cdot \begin{bmatrix} (I * H_x^S)(u,v) \\ (I * H_y^S)(u,v) \end{bmatrix} \tag{6.12}$$

for the *Sobel* operator.

---

[2] In Eqn. (6.9), $*$ is the linear convolution operator (see Sec. 5.3.1).

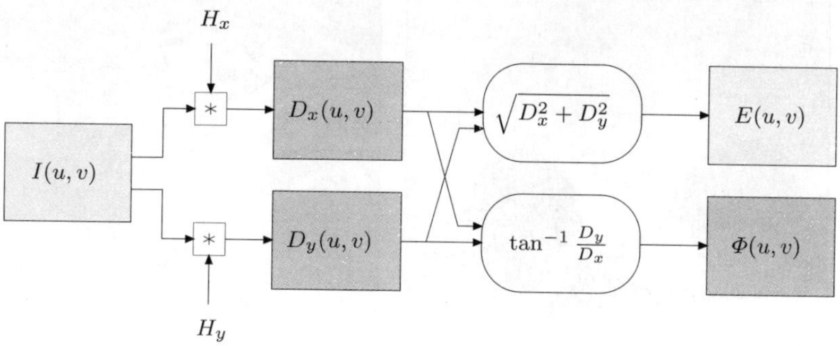

Figure 6.5 Typical process of gradient-based edge extraction. The two linear gradient filters H_x and H_y produce two gradient images, D_x and D_y, respectively. They are used to compute the edge strength E and orientation Φ for each image position (u, v).

Edge strength and orientation

In the following, we denote the scaled filter results (obtained with either the Prewitt or Sobel operator) as

$$D_x = H_x * I \qquad \text{and} \qquad D_y = H_y * I.$$

In both cases, the local edge strength $E(u, v)$ is defined as the gradient magnitude

$$E(u,v) = \sqrt{\big(D_x(u,v)\big)^2 + \big(D_y(u,v)\big)^2}, \tag{6.13}$$

and the local edge orientation angle $\Phi(u, v)$ is[3]

$$\Phi(u,v) = \tan^{-1}\Big(\frac{D_y(u,v)}{D_x(u,v)}\Big) = \mathrm{Arctan}\big(D_y(u,v), D_x(u,v)\big). \tag{6.14}$$

The whole process of extracting the edge magnitude and orientation is summarized in Fig. 6.5. First, the original image I is independently convolved with the two gradient filters H_x and H_y, and subsequently the edge strength E and orientation Φ are computed from the filter results. Figure 6.6 shows the edge strength and orientation for two test images, obtained with the Sobel filters in Eqn. (6.10).

The estimate of the edge orientation based on the original Prewitt and Sobel filters is relatively inaccurate, and improved versions of the Sobel filters were proposed in [24, p. 353] to minimize the orientation errors:

$$H_x^{S'} = \frac{1}{32}\begin{bmatrix} -3 & 0 & 3 \\ -10 & \mathbf{0} & 10 \\ -3 & 0 & 3 \end{bmatrix} \quad \text{and} \quad H_y^{S'} = \frac{1}{32}\begin{bmatrix} -3 & -10 & -3 \\ 0 & \mathbf{0} & 0 \\ 3 & 10 & 3 \end{bmatrix}. \tag{6.15}$$

[3] See the hints in Appendix B.1.6 for computing the inverse tangent $\tan^{-1}(y/x)$ with the $\mathrm{Arctan}(y, x)$ function.

(a)

(b)

(c)

(d)

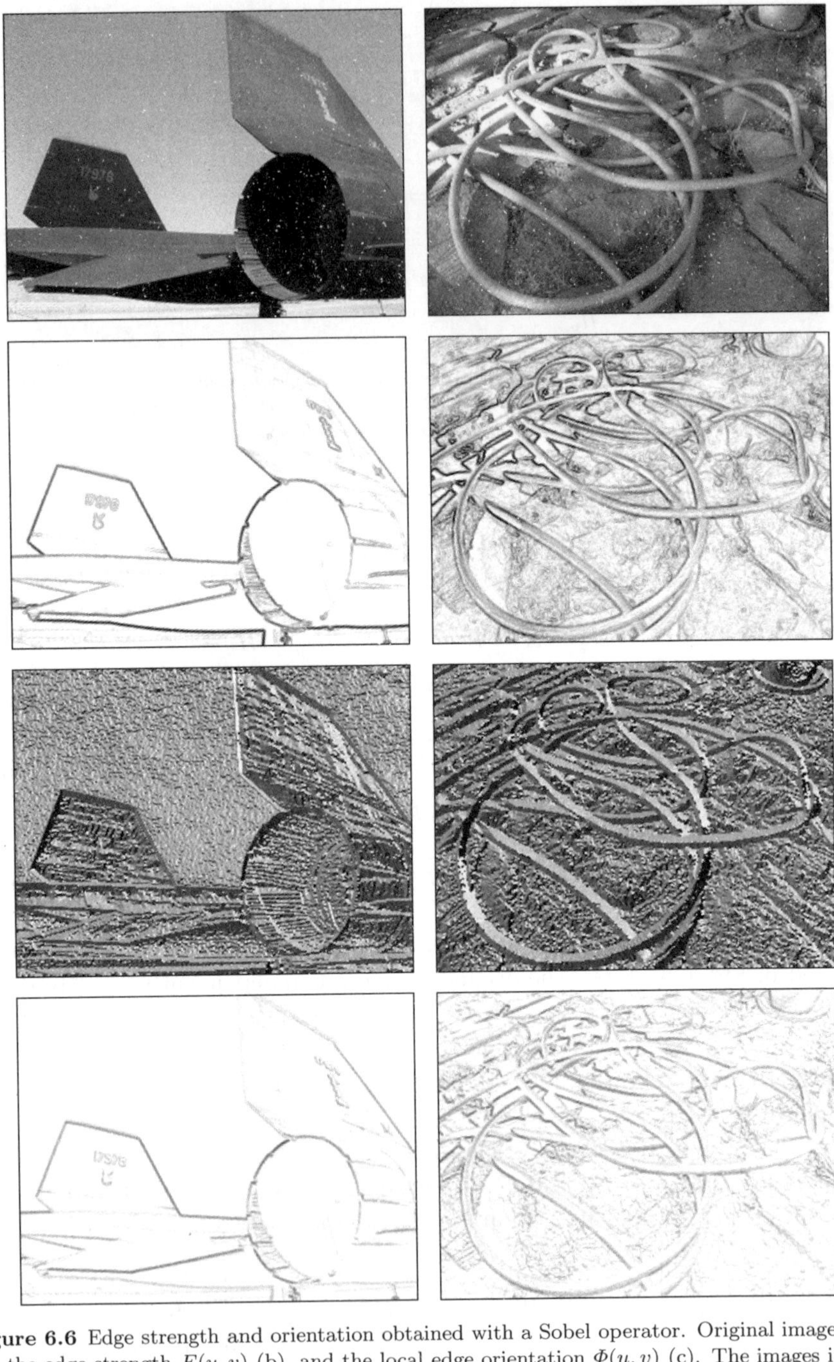

Figure 6.6 Edge strength and orientation obtained with a Sobel operator. Original images (a), the edge strength $E(u, v)$ (b), and the local edge orientation $\Phi(u, v)$ (c). The images in (d) show the orientation angles coded as color hues, with the edge strength controlling the color saturation (see Sec. 8.2.3 for the corresponding definitions).

These edge operators are frequently used because of their good results (see also Fig. 6.11) and simple implementation. The Sobel operator, in particular, is available in many image-processing tools and software packages (including ImageJ).

6.3.2 Roberts Operator

As one of the simplest and oldest edge finders, the Roberts operator [36] today is mainly of historical interest. It employs two extremely small filters of size 2×2 for estimating the directional gradient along the image diagonals:

$$H_1^R = \begin{bmatrix} 0 & \mathbf{1} \\ -1 & 0 \end{bmatrix} \quad \text{and} \quad H_2^R = \begin{bmatrix} -1 & 0 \\ 0 & \mathbf{1} \end{bmatrix}. \tag{6.16}$$

These filters naturally respond to diagonal edges but are not highly selective to orientation; i. e., both filters show strong results over a relatively wide range of angles (Fig. 6.7). The local edge strength is computed by measuring the length of the resulting 2D vector, similar to the gradient computation but with its components rotated 45° (Fig. 6.8).

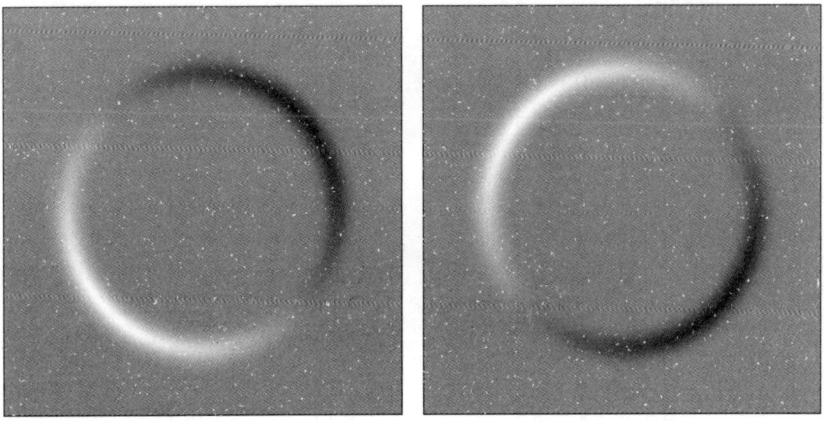

$D_1 = I * H_1^R$ $\qquad$ $D_2 = I * H_2^R$

Figure 6.7 Diagonal gradient components produced by the two Roberts filters.

6.3.3 Compass Operators

The design of linear edge filters involves a trade-off: the stronger a filter responds to edge-like structures, the more sensitive it is to orientation. In other words, filters that are orientation-insensitive tend to respond to nonedge structures, while the most discriminating edge filters only respond to edges in a

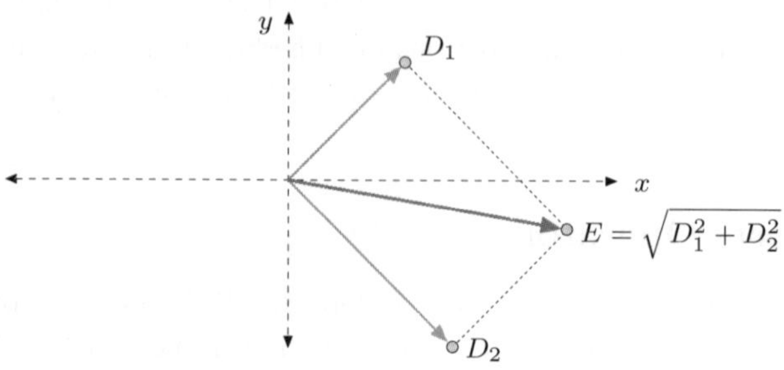

Figure 6.8 Definition of edge strength for the Roberts operator. The edge strength $E(u, v)$ corresponds to the length of the vector obtained by adding the two orthogonal gradient components (filter results) $D_1(u, v)$ and $D_2(u, v)$.

narrow range of orientations. One solution is to use not only a single pair of relatively "wide" filters for two directions (such as the Prewitt and the simple Sobel operator discussed above) but a larger set of filters with narrowly spaced orientations. A classic example is the extended Sobel operator, which employs the following eight filters with orientations spaced at 45°:

$$H_0^S = \begin{bmatrix} -1 & 0 & 1 \\ -2 & 0 & 2 \\ -1 & 0 & 1 \end{bmatrix} \qquad H_4^S = \begin{bmatrix} 1 & 0 & -1 \\ 2 & 0 & -2 \\ 1 & 0 & -1 \end{bmatrix}, \tag{6.17}$$

$$H_1^S = \begin{bmatrix} -2 & -1 & 0 \\ -1 & 0 & 1 \\ 0 & 1 & 2 \end{bmatrix} \qquad H_5^S = \begin{bmatrix} 2 & 1 & 0 \\ 1 & 0 & -1 \\ 0 & -1 & -2 \end{bmatrix}, \tag{6.18}$$

$$H_2^S = \begin{bmatrix} -1 & -2 & -1 \\ 0 & 0 & 0 \\ 1 & 2 & 1 \end{bmatrix} \qquad H_6^S = \begin{bmatrix} 1 & 2 & 1 \\ 0 & 0 & 0 \\ -1 & -2 & -1 \end{bmatrix}, \tag{6.19}$$

$$H_3^S = \begin{bmatrix} 0 & -1 & -2 \\ 1 & 0 & -1 \\ 2 & 1 & 0 \end{bmatrix} \qquad H_7^S = \begin{bmatrix} 0 & 1 & 2 \\ -1 & 0 & 1 \\ -2 & -1 & 0 \end{bmatrix}. \tag{6.20}$$

Only the results of four of the eight filters $H_0^S, H_1^S, \ldots H_7^S$ above must actually be computed since the remaining four are identical except for the reversed sign. For example, from the fact that $H_4^S = -H_0^S$ and the convolution being linear (Eqn. (5.18)), it follows that

$$I * H_4^S = I * -H_0^S = -(I * H_0^S); \tag{6.21}$$

i. e., the result for filter H_4^S is simply the negative result for filter H_0^S. The directional outputs $D_0, D_1, \ldots D_7$ for the eight Sobel filters can thus be computed as follows:

$$\begin{array}{llll} D_0 \leftarrow I * H_0^S & D_1 \leftarrow I * H_1^S & D_2 \leftarrow I * H_2^S & D_3 \leftarrow I * H_3^S \\ D_4 \leftarrow -D_0 & D_5 \leftarrow -D_1 & D_6 \leftarrow -D_2 & D_7 \leftarrow -D_3. \end{array} \tag{6.22}$$

The edge strength E^S at position (u, v) is defined as the maximum of the eight filter outputs; i. e.,

$$\begin{aligned} E^S(u,v) &\triangleq \max\bigl(D_0(u,v), D_1(u,v), D_2(u,v), D_3(u,v), D_4(u,v), \ldots D_7(u,v)\bigr) \\ &= \max\bigl(|D_0(u,v)|, |D_1(u,v)|, |D_2(u,v)|, |D_3(u,v)|\bigr), \end{aligned} \tag{6.23}$$

and the strongest-responding filter also determines the local edge orientation as

$$\Phi^S(u,v) \triangleq \frac{\pi}{4} j, \qquad \text{with } j = \underset{0 \le i \le 7}{\operatorname{argmax}}\; D_i(u,v). \tag{6.24}$$

Another classic compass operator is the one proposed by *Kirsch* [27], which is also based on eight directional filters with the following kernels:

$$H_0^K = \begin{bmatrix} -5 & 3 & 3 \\ -5 & 0 & 3 \\ -5 & 3 & 3 \end{bmatrix} \qquad H_4^K = \begin{bmatrix} 3 & 3 & -5 \\ 3 & 0 & -5 \\ 3 & 3 & -5 \end{bmatrix}, \tag{6.25}$$

$$H_1^K = \begin{bmatrix} -5 & -5 & 3 \\ -5 & 0 & 3 \\ 3 & 3 & 3 \end{bmatrix} \qquad H_5^K = \begin{bmatrix} 3 & 3 & 3 \\ 3 & 0 & -5 \\ 3 & -5 & -5 \end{bmatrix}, \tag{6.26}$$

$$H_2^K = \begin{bmatrix} -5 & -5 & -5 \\ 3 & 0 & 0 \\ 3 & 3 & 3 \end{bmatrix} \qquad H_6^K = \begin{bmatrix} 3 & 3 & 3 \\ 3 & 0 & 3 \\ -5 & -5 & -5 \end{bmatrix}, \tag{6.27}$$

$$H_3^K = \begin{bmatrix} 3 & -5 & -5 \\ 3 & 0 & -5 \\ 3 & 3 & 3 \end{bmatrix} \qquad H_7^K = \begin{bmatrix} 3 & 3 & 3 \\ -5 & 0 & 3 \\ -5 & -5 & 3 \end{bmatrix}. \tag{6.28}$$

Again, because of the symmetries, only four of the eight filters need to be applied and the results may be combined in the same way as described above for the extended Sobel operator. In practice, this and other "compass operators" show only minor benefits over the simpler operators described earlier, including the small advantage of not requiring the computation of square roots (which is considered a relatively "expensive" operation).

6.3.4 Edge Operators in ImageJ

The current version of ImageJ implements the Sobel operator (as described in Eqn. (6.10)) for practically any type of image. It can be invoked via the

Process→Find Edges

menu and is also available through the method `void findEdges()` for objects of type `ImageProcessor`.

6.4 Other Edge Operators

One problem with edge operators based on first derivatives (as described in the previous section) is that each resulting edge is as wide as the underlying intensity transition and thus edges may be difficult to localize precisely. An alternative class of edge operators makes use of the second derivatives of the image function, including some popular modern edge operators that also address the problem of edges appearing at various levels of scale. These issues are briefly discussed in the following.

6.4.1 Edge Detection Based on Second Derivatives

The second derivative of a function measures its local curvature. The idea is that edges can be found at zero positions or—even better—at the zero crossings of the second derivatives of the image function, as illustrated in Fig. 6.9 for the one-dimensional case. Since second derivatives generally tend to amplify image noise, some sort of presmoothing is usually applied with suitable low-pass filters.

A popular example is the "Laplacian-of-Gaussian" (LoG) operator [29], which combines gaussian smoothing and computing the second derivatives (see the *Laplace Filter* in Sec. 6.6.1) into a single linear filter. The example in Fig. 6.11 shows that the edges produced by the LoG operator are more precisely localized than the ones delivered by the Prewitt and Sobel operators, and the amount of "clutter" is comparably small. Details about the LoG operator and a comprehensive survey of common edge operators can be found in [37, Ch. 4] and [31].

6.4.2 Edges at Different Scales

Unfortunately, the results of the simple edge operators we have discussed so far often deviate from what we as humans perceive as important edges. The two main reasons for this are:

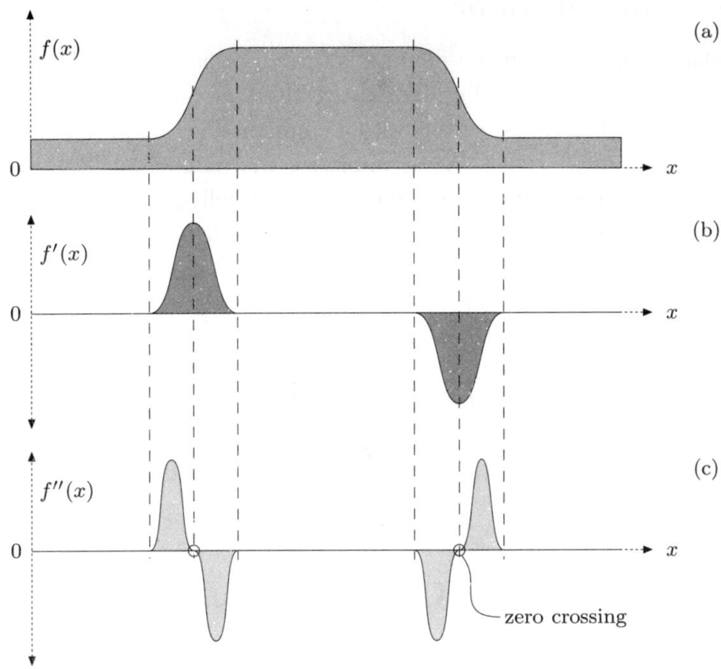

Figure 6.9 Principle of edge detection with the second derivative: original function (a), first derivative (b), and second derivative (c). Edge points are located where the second derivative crosses through zero and the first derivative has a high magnitude.

- First, edge operators only respond to local intensity differences, while our visual system is able to extend edges across areas of minimal or vanishing contrast.
- Second, edges exist not at a single fixed resolution or at a certain scale but over a whole range of different scales.

Typical small edge operators, such as the Sobel operator, can only respond to intensity differences that occur within their 3×3 pixel filter regions. To recognize edge-like events over a greater horizon, we would either need larger edge operators (with correspondingly large filters) or use the original (small) operators on reduced (i. e., scaled) images. This is the principal idea of "multiresolution" techniques (also referred to as "hierarchical" or "pyramid" techniques), which have traditionally been used in many image-processing applications [7, 28]. In the context of edge detection, this typically amounts to detecting edges at various scale levels first and then deciding which edge (if any) at which scale level is dominant at each image position.

6.4.3 Canny Operator

A popular example for such a method is the edge operator by Canny [8], which employs a set of relatively large, oriented filters at multiple image resolutions and merges the individual results into a common edge map. The method tries to reach three main goals: (a) to minimize the number of false edge points, (b) achieve good localization of edges, and (c) deliver only a single mark on each edge. At its core, the Canny "filter" is a gradient method (based on first derivatives; see Sec. 6.2), but it uses the zero crossings of second derivatives for precise edge localization. Frequently, however, only a single-scale implementation of the algorithm with an adjustable filter radius (smoothing parameter σ) is used, which is nevertheless superior to most of the simple edge operators (see Figs. 6.10 and 6.11). Thus, even in its basic (single-scale) form, the Canny operator is often preferred over other edge detection methods. A more detailed description of the algorithm and a Java implementation can be found, for example, in [12, Ch. 7].

6.5 From Edges to Contours

Whatever method is used for edge detection, the result is usually a continuous value for the edge strength for each image position and possibly also the angle of local edge orientation. How can this information be used, for example, to find larger image structures and contours of objects in particular?

6.5.1 Contour Following

The idea of tracing contours sequentially along the discovered edge points is not uncommon and appears quite simple in principle. Starting from an image point with high edge strength, the edge is followed iteratively in both directions until the two traces meet and a closed contour is formed. Unfortunately, there are several obstacles that make this task more difficult than it seems at first, including the following:

- Edges may end in regions of vanishing intensity gradient.
- Crossing edges lead to ambiguities.
- Contours may branch into several directions.

Because of these problems, contour following usually is not applied to original images or continuous-valued edge images except in very simple situations, such as when there is a clear separation between objects (foreground) and the background. Tracing contours in binary images is much simpler, of course (see Vol. 2 [6, Ch. 2]).

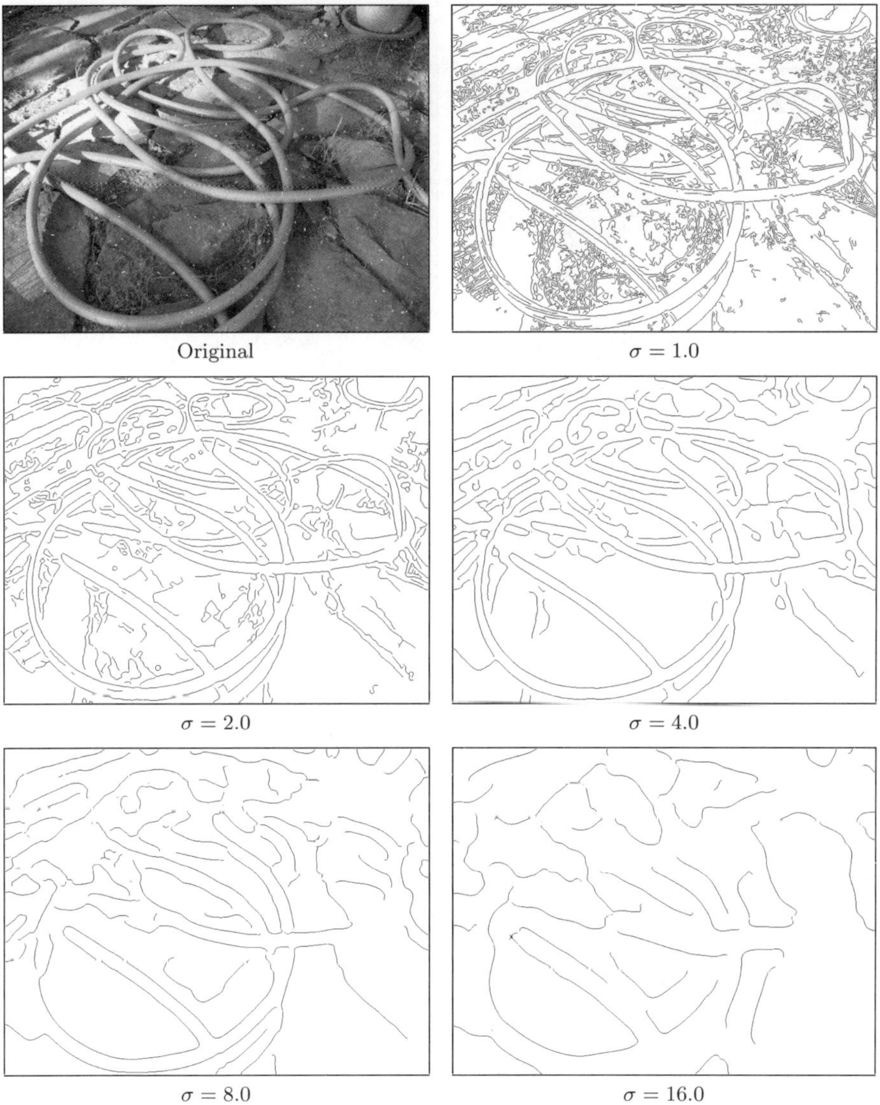

Figure 6.10 Canny edge operator. Resulting edge maps for different settings of the smoothing (scale) parameter σ.

6.5.2 Edge Maps

In many situations, the next step after edge enhancement (by some edge operator) is the selection of edge points, a binary decision whether an image pixel is an edge point or not. The simplest method is to apply a *threshold* operation to the edge strength delivered by the edge operator using either a fixed or

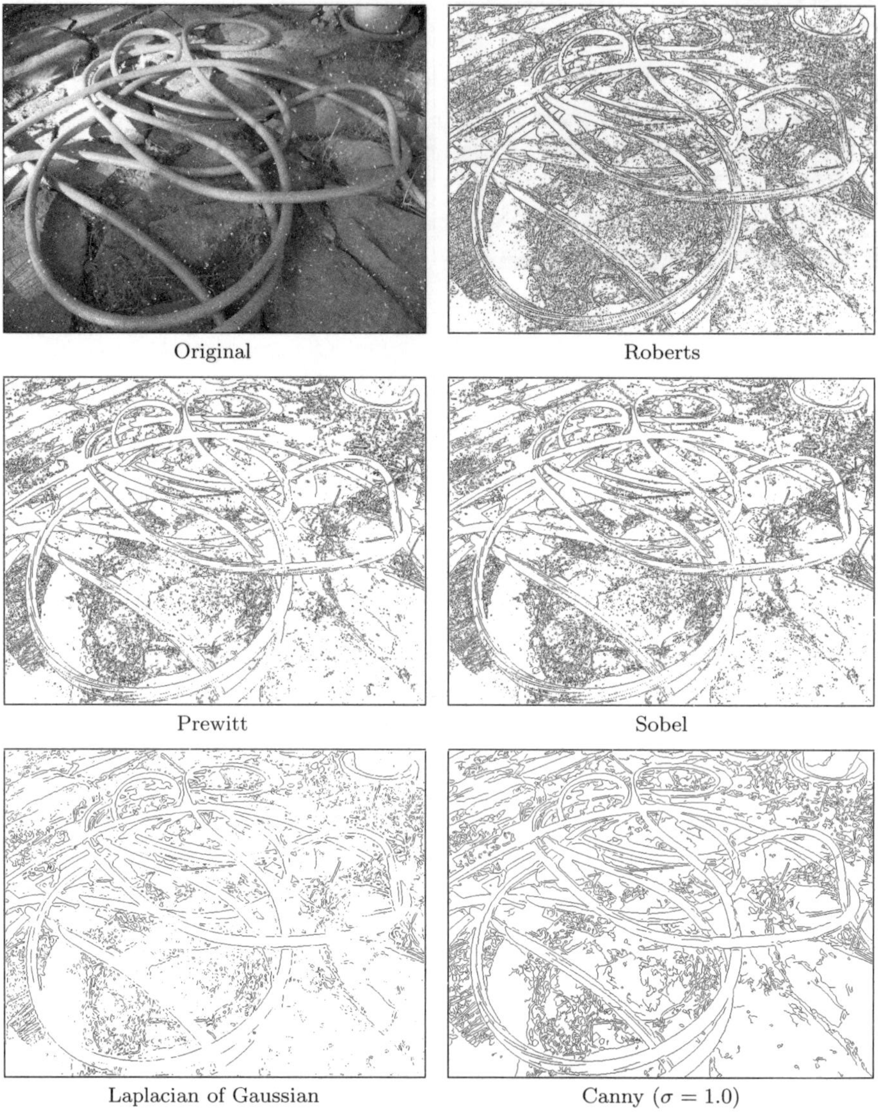

Figure 6.11 Comparison of various edge operators. Important criteria for the quality of edge results are the amount of "clutter" (irrelevant edge elements) and the connectedness of dominant edges. The Roberts operator responds to very small edge structures because of the small size of its filters. The similarity of the Prewitt and Sobel operators is manifested in the corresponding results. The edge map produced by the Canny operator is substantially cleaner than those of the simpler operators, even for a fixed and relatively small scale value σ.

adaptive threshold value, which results in a binary edge image or "edge map".

In practice, edge maps hardly ever contain perfect contours but instead many small, unconnected contour fragments, interrupted at positions of insufficient edge strength. After thresholding, the empty positions of course contain no edge information at all that could possibly be used in a subsequent step, such as for linking adjacent edge segments. Despite this weakness, global thresholding is often used at this point because of its simplicity, and some common postprocessing methods, such as the Hough transform (see Vol. 2 [6, Ch. 3]), can cope well with incomplete edge maps.

6.6 Edge Sharpening

Making images look sharper is a frequent task, such as to make up for a lack of sharpness after scanning or scaling an image or to precompensate for a subsequent loss of sharpness in the course of printing or displaying an image. The common approach to image sharpening is to amplify the high-frequency image components, which are mainly responsible for the perceived sharpness of an image and for which the strongest occur at rapid intensity transitions. In the following, we describe two methods for artificial image sharpening that are based on techniques similar to edge detection and thus fit well in this chapter.

6.6.1 Edge Sharpening with the Laplace Filter

A common method for localizing rapid intensity changes are filters based on the second derivatives of the image function. Figure 6.12 illustrates this idea on a one-dimensional, continuous function $f(x)$. The second derivative $f''(x)$ of the step function shows a positive pulse at the lower end of the transition and a negative pulse at the upper end. The edge is sharpened by subtracting a certain fraction w of the second derivative $f''(x)$ from the original function $f(x)$,

$$\check{f}(x) = f(x) - w \cdot f''(x). \tag{6.29}$$

Depending upon the weight factor $w \geq 0$, the expression in Eqn. (6.29) causes the intensity function to overshoot at both sides of an edge, thus exaggerating edges and increasing the perceived sharpness.

Laplace operator

Sharpening of a two-dimensional function can be accomplished with the second derivatives in the horizontal and vertical directions combined by the so-called Laplace operator. The Laplace operator ∇^2 of a two-dimensional function

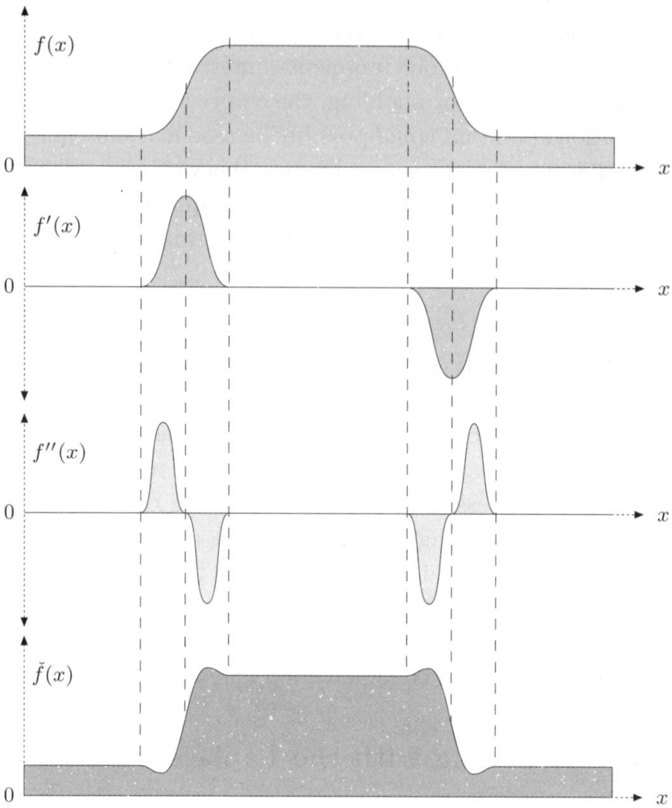

Figure 6.12 Edge sharpening with the second derivative. The original intensity function $f(x)$, first derivative $f'(x)$, second derivative $f''(x)$, and sharpened intensity function $\check{f}(x) = f(x) - w \cdot f''(x)$ are shown.

$f(x, y)$ is defined as the sum of the second partial derivatives along the x and y directions:

$$(\nabla^2 f)(x, y) = \frac{\partial^2 f}{\partial^2 x}(x, y) + \frac{\partial^2 f}{\partial^2 y}(x, y). \tag{6.30}$$

Similar to the first derivatives (see Sec. 6.2.2), the second derivatives of a discrete image function can also be estimated with a set of simple linear filters. Again, several versions, have been proposed. For example, the two one-dimensional filters

$$\frac{\partial^2 f}{\partial^2 x} \equiv H_x^L = \begin{bmatrix} 1 & \mathbf{-2} & 1 \end{bmatrix} \quad \text{and} \quad \frac{\partial^2 f}{\partial^2 y} \equiv H_y^L = \begin{bmatrix} 1 \\ \mathbf{-2} \\ 1 \end{bmatrix} \tag{6.31}$$

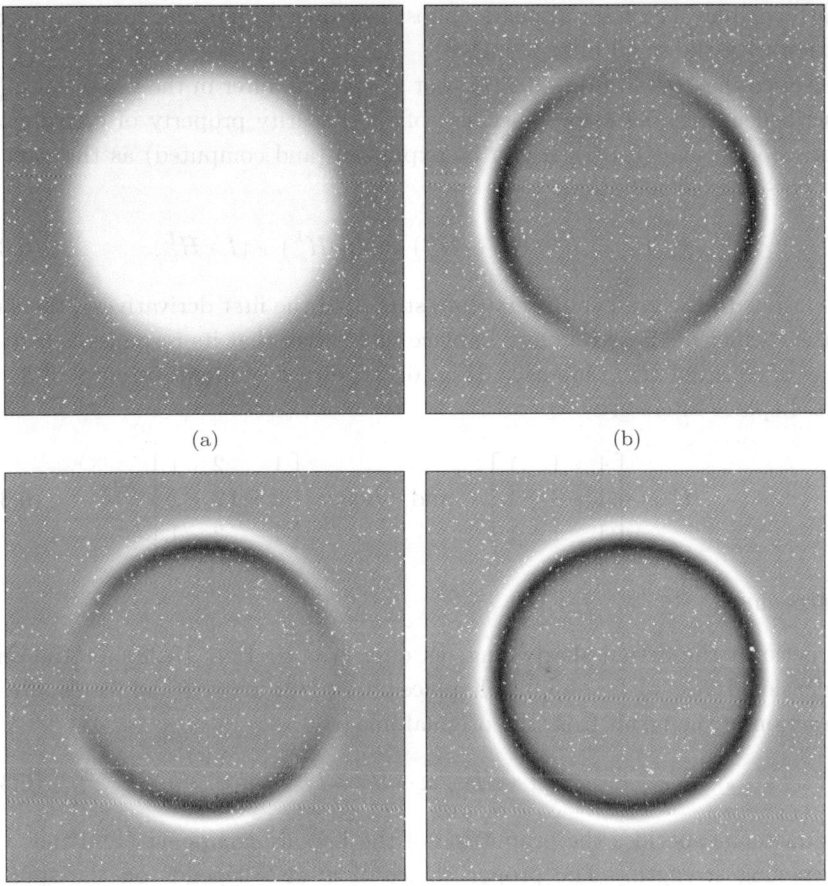

(a) (b) (c) (d)

Figure 6.13 Results of Laplace filter H^L: synthetic test image I (a), second partial derivative $\partial^2 I/\partial^2 u$ in the horizontal direction (b), second partial derivative $\partial^2 I/\partial^2 v$ in the vertical direction (c), and Laplace filter $\nabla^2 I(u,v)$ (d). Intensities in (b–d) are scaled such that maximally negative and positive values are shown as black and white, respectively, and zero values are gray.

for estimating the second derivatives along the x and y directions, respectively, combine to make the two-dimensional Laplace filter

$$H^L = H_x^L + H_y^L = \begin{bmatrix} 0 & 1 & 0 \\ 1 & -4 & 1 \\ 0 & 1 & 0 \end{bmatrix}. \tag{6.32}$$

Figure 6.13 shows an example of applying the Laplace filter H^L to a grayscale image, where the pairs of positive-negative peaks at both sides of each edge are clearly visible. The filter appears almost isotropic despite the coarse approximation with the small filter kernels.

Notice that H^L in Eqn. (6.32) is not a *separable* filter in the usual sense (as described in Sec. 5.3.3) but, because of the linearity property of convolution (Eqns. (5.17) and (5.19)), it can be expressed (and computed) as the *sum* of two one-dimensional filters,

$$I * H^L = I * (H_x^L + H_y^L) = (I * H_x^L) + (I * H_y^L). \tag{6.33}$$

Analogous to the gradient filters (for estimating the first derivatives), the sum of the coefficients is zero in any Laplace filter, such that its response is zero in areas of constant (flat) intensity (Fig. 6.13). Other common variants of 3×3 pixel Laplace filters are

$$H_8^L = \begin{bmatrix} 1 & 1 & 1 \\ 1 & \mathbf{-8} & 1 \\ 1 & 1 & 1 \end{bmatrix} \quad \text{and} \quad H_{12}^L = \begin{bmatrix} 1 & 2 & 1 \\ 2 & \mathbf{-12} & 2 \\ 1 & 2 & 1 \end{bmatrix}. \tag{6.34}$$

Sharpening

To perform the actual sharpening, as described by Eqn. (6.29) for the one-dimensional case, we first apply a Laplace filter to the image I and then subtract a fraction of the result from the original image,

$$\check{I} \leftarrow I - w \cdot (H^L * I). \tag{6.35}$$

The factor w specifies the proportion of the Laplace component and thus the sharpening strength. The proper choice of w also depends on the specific Laplace filter used in Eqn. (6.35) since none of the filters above is normalized.

Figure 6.13 shows the result of applying a Laplace filter (with the kernel given in Eqn. (6.32)) to a synthetic test image where the pairs of positive/negative peaks at both sides of each edge are clearly visible. The filter appears almost isotropic despite the coarse approximation with the small filter kernels. The application to a real grayscale image using the filter H^L (Eqn. (6.32)) and $w = 1.0$ is shown in Fig. 6.14.

As we can expect from second-order derivatives, the Laplace filter is fairly sensitive to image noise, which can be reduced (as is commonly done in edge detection with first derivatives) by previous smoothing such as with a Gaussian filter (see also Sec. 6.4.1).

6.6.2 Unsharp Masking

"Unsharp masking" (USM) is a technique for edge sharpening that is particularly popular in astronomy, digital printing, and many other areas of image

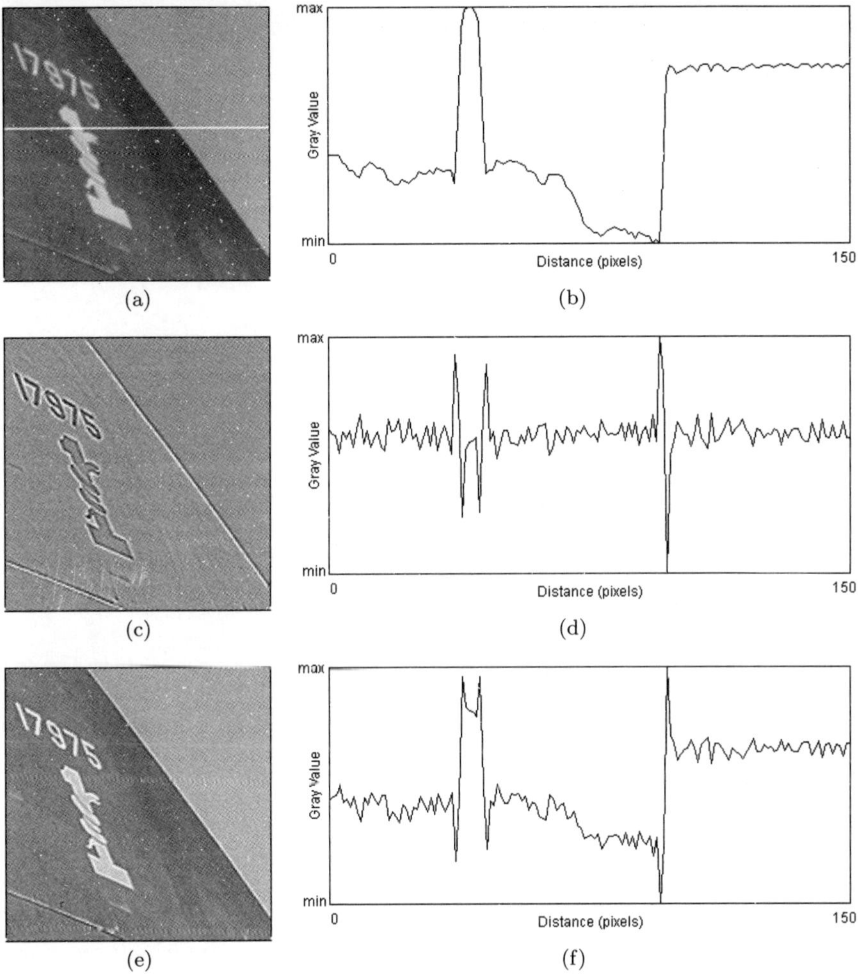

Figure 6.14 Edge sharpening with the Laplace filter: original image with a horizontal profile taken from the marked line (a, b), result of Laplace filter H^L (c, d), and sharpened image (e, f).

processing. The term originates from classical photography, where the sharpness of an image was optically enhanced by combining it with a smoothed ("unsharp") copy. This process is in principle the same for digital images.

Process

The first step in the USM filter is to subtract a smoothed version of the image from the original, which enhances the edges. The result is called the "mask". In analog photography, the required smoothing was achieved by simply defocusing the lens. Subsequently, the mask is again added to the original, such that the edges in the image are sharpened. In summary, the steps involved in USM filtering are:

1. The mask M is generated by subtracting a smoothed version of the image I from the original,

$$M \leftarrow I - (I * \tilde{H}) = I - \tilde{I}, \tag{6.36}$$

where the kernel $\tilde{H}$ of the smoothing filter is assumed to be normalized (see Sec. 5.2.5).

2. To obtain the sharpened image $\check{I}$, the mask M is added to the original image I, weighted by the factor a, which controls the amount of sharpening,

$$\check{I} \leftarrow I + a \cdot M, \tag{6.37}$$

and thus (substituting from Eqn. (6.36))

$$\check{I} \leftarrow I + a \cdot (I - \tilde{I}) = (1 + a) \cdot I - a \cdot \tilde{I}. \tag{6.38}$$

Smoothing filter

In principle, any smoothing filter could be used for the kernel $\tilde{H}$ in Eqn. (6.36), but Gaussian filters $H^{G,\sigma}$ with variable radius σ are most common (see also Sec. 5.2.7). Typical parameter values are 1 to 20 for σ and 0.2 to 4.0 (equivalent to 20% to 400%) for the sharpening factor a. Figure 6.15 shows two examples of USM filters using Gaussian smoothing filters with different radii σ.

Extensions

The advantages of the USM filter over the Laplace filter are a reduced noise sensitivity due to the involved smoothing and improved controllability through the parameters σ (spatial extent) and a (sharpening strength).

Of course the USM filter responds not only to real edges but to some extent to any intensity transition and thus potentially increases any visible noise in continuous image regions. Some implementations (e. g., Adobe Photoshop) therefore provide an additional *threshold* parameter t_c to specify the *minimum local contrast* required to perform edge sharpening. Sharpening is only applied if the local contrast at position (u, v), expressed for example by the gradient magnitude $|\nabla I|$ (Eqn. (6.5)), is greater than that threshold. Otherwise, that pixel remains unmodified:

$$\check{I}(u,v) \leftarrow \begin{cases} I(u,v) + a \cdot M(u,v) & \text{for } |\nabla I|(u,v) \geq t_c \\ I(u,v) & \text{otherwise.} \end{cases} \tag{6.39}$$

(a) Original

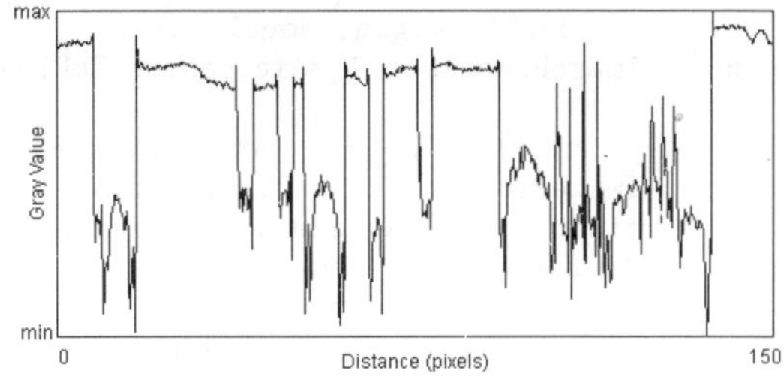

(b)

(c) $\sigma = 2.5$

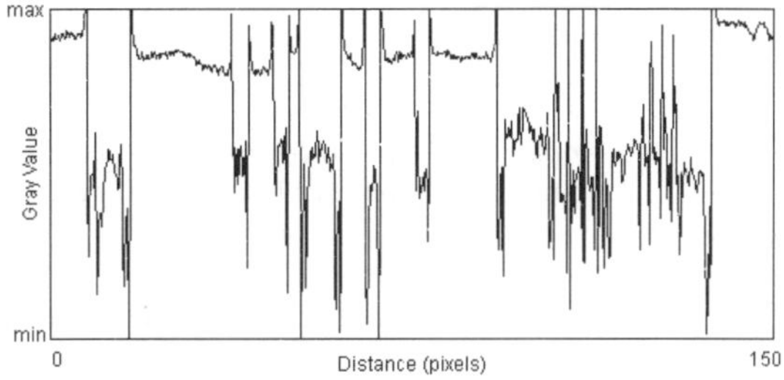

(d)

(e) $\sigma = 10.0$

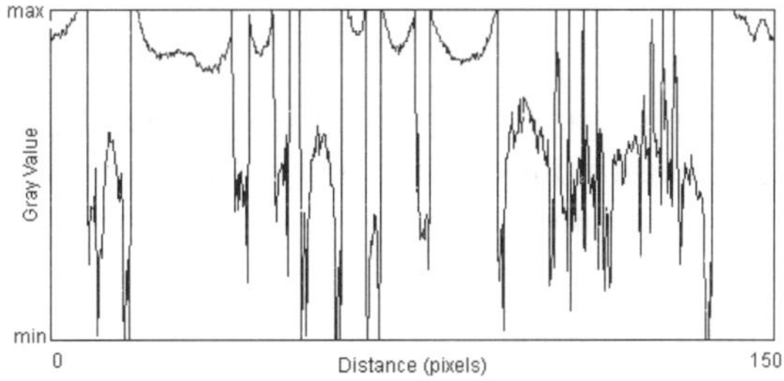

(f)

(g) Original

(h) $\sigma = 2.5$

(i) $\sigma = 10.0$

Figure 6.15 USM filters with varying smoothing radii σ. Original image (a) and the intensity profile along the marked image line (b); results of USM filtering with Gaussian smoothing radius $\sigma = 2.5$ (c, d) and $\sigma = 10.0$ (e, f); enlarged image detail (g–i). The value of the sharpening factor a is 1.0 (100%).

```
1  public void unsharpMask(ImageProcessor ip,
2          double sigma, double a) {
3    ImageProcessor I = ip.convertToFloat(); // I
4
5    // create a blurred version of the image
6    ImageProcessor J = I.duplicate();         // Ĩ
7    float[] H = GaussKernel1d.create(sigma);  // see Prog. 5.4
8    Convolver cv = new Convolver();
9    cv.setNormalize(true);
10   cv.convolve(J, H, 1, H.length);
11   cv.convolve(J, H, H.length, 1);
12
13   I.multiply(1+a);      // I ← (1 + a) · I
14   J.multiply(a);        // Ĩ ← a · Ĩ
15   I.copyBits(J,0,0,Blitter.SUBTRACT); // Ĩ ← (1 + a) · I − a · Ĩ
16
17   //copy result back into original byte image
18   ip.insert(I.convertToByte(false), 0, 0);
19 }
```

Program 6.1 Unsharp masking (Java implementation). First the original image is converted to a `FloatProcessor` object `I` (I) in line 3, which is duplicated to hold the blurred image `J` ($\tilde{I}$) in line 6. The method `makeGaussKernel1d()`, defined in Prog. 5.4, is used to create the 1D Gaussian filter kernel applied in the horizontal and vertical directions (lines 10–11). The remaining computations follow Eqn. (6.38).

Different from the original USM filter (Eqn. (6.37)), this extended version is no longer a linear filter. On color images, the USM filter is usually applied to all color channels with identical parameter settings.

Implementation

The USM filter is available in virtually any image-processing software and, due to its simplicity and flexibility, has become an indispensable tool for many professional users. In ImageJ, the USM filter is implemented by the plugin class `ij.plugin.filter.UnsharpMask`, which can be invoked through the menu

Process→Filter→Unsharp Mask...

ImageJ's `UnsharpMask` implementation uses the class `GaussianBlur` for the required smoothing operation. The implementation shown in Prog. 6.1 follows the definition in Eqn. (6.38) and uses relatively large filter kernels that are created with the method `makeGaussKernel1d()`, as defined in Prog. 5.4.

Laplace versus USM filter

A closer look at these two methods reveals that sharpening with the Laplace filter (Sec. 6.6.1) can be viewed as a special case of the USM filter. If the

Laplace filter in Eqn. (6.32) is decomposed as

$$H^L = \begin{bmatrix} 0 & 1 & 0 \\ 1 & \mathbf{-4} & 1 \\ 0 & 1 & 0 \end{bmatrix} = \begin{bmatrix} 0 & 1 & 0 \\ 1 & \mathbf{1} & 1 \\ 0 & 1 & 0 \end{bmatrix} - 5 \begin{bmatrix} 0 & 0 & 0 \\ 0 & \mathbf{1} & 0 \\ 0 & 0 & 0 \end{bmatrix} \tag{6.40}$$

$$= 5 \left(\frac{1}{5} \begin{bmatrix} 0 & 1 & 0 \\ 1 & \mathbf{1} & 1 \\ 0 & 1 & 0 \end{bmatrix} - \begin{bmatrix} 0 & 0 & 0 \\ 0 & \mathbf{1} & 0 \\ 0 & 0 & 0 \end{bmatrix} \right) = 5\,(\tilde{H} - \delta),$$

one can see that H^L consists of a simple 3×3 pixel smoothing filter $\tilde{H}$ minus the impulse function δ. Laplace sharpening with the weight factor w as defined in Eqn. (6.35) can therefore (by a little manipulation) be expressed as

$$\begin{aligned} \check{I}_L &\leftarrow I - w \cdot (H^L * I) \;=\; I - w \cdot \big(5(\tilde{H}^L - \delta) * I\big) \\ &= I - 5w \cdot (\tilde{H}^L * I - I) \;=\; I + 5w \cdot (I - \tilde{H}^L * I) \\ &= I + 5w \cdot M^L; \end{aligned} \tag{6.41}$$

i. e., in the form of a USM filter $\check{I} \leftarrow I + a \cdot M$ (Eqn. (6.37)). Laplacian sharpening is thus a special case of a USM filter with the mask $M = M^L = (I - \tilde{H}^L * I)$, the specific smoothing filter

$$\tilde{H} = \tilde{H}^L = \frac{1}{5} \begin{bmatrix} 0 & 1 & 0 \\ 1 & \mathbf{1} & 1 \\ 0 & 1 & 0 \end{bmatrix},$$

and the sharpening factor $a = 5w$.

6.7 Exercises

Exercise 6.1

Compute manually the gradient and the Laplacian for the following image $I(u,v)$ by using the approximations in Eqns. (6.2) and (6.32), respectively:

$$I(u,v) = \begin{bmatrix} 14 & 10 & 19 & 16 & 14 & 12 \\ 18 & 9 & 11 & 12 & 10 & 19 \\ 9 & 14 & 15 & 26 & 13 & 6 \\ 21 & 27 & 17 & 17 & 19 & 16 \\ 11 & 18 & 18 & 19 & 16 & 14 \\ 16 & 10 & 13 & 7 & 22 & 21 \end{bmatrix}.$$

Exercise 6.2

Implement the Sobel edge operator as defined in Eqn. (6.10) (and illustrated in Fig. 6.5) as an ImageJ plugin. The plugin should generate two new images for the edge magnitude $E(u, v)$ and the edge orientation $\Phi(u, v)$. Come up with a suitable way to display the edge orientation.

Exercise 6.3

Express the Sobel operator in x/y-separable form analogous to the decomposition of the Prewitt operator in Eqn. (6.9).

Exercise 6.4

Implement the Kirsch operator (Eqn. (6.25)) analogous to the two-directional Sobel operator in Exercise 6.2 and compare the results from both methods, particularly the edge orientation estimates.

Exercise 6.5

Devise and implement a compass edge operator with more than 8 (16?) differently oriented filters.

Exercise 6.6

Compare the results of the unsharp masking filters in ImageJ and Adobe Photoshop using a suitable test image. How should the parameters for σ (*radius*) and a (*weight*) be defined in both implementations to obtain similar results?

7 Morphological Filters

In the discussion of the median filter in Ch. 5 (Sec. 5.4.2), we noticed that this type of filter can somehow alter two-dimensional image structures. Figure 7.1 illustrates once more how corners are rounded off, holes of a certain size are filled, and small structures, such as single dots or thin lines, are removed. The median filter thus responds selectively to the local shape of image structures, a property that might be useful for other purposes if it can be applied not just randomly but in a controlled fashion. Altering the local structure in a predictable way is exactly what "morphological" filters can do, which we focus on in this chapter.

In their original form, morphological filters are aimed at binary images,

(a) (b) (c)

Figure 7.1 Median filter applied to a binary image: original image (a) and results from a 3×3 pixel median filter (b) and a 5×5 pixel median filter (c).

W. Burger, M.J. Burge, *Principles of Digital Image Processing*, Undergraduate Topics in Computer Science, DOI 10.1007/978-1-84800-191-6_7, © Springer-Verlag London Limited, 2009

images with only two possible pixel values, 0 and 1 or *black* and *white*, respectively. Binary images are found in many places, in particular in digital printing, document transmission (FAX) and storage, or as selection masks in image and video editing. Binary images can be obtained from grayscale images by simple thresholding (see Sec. 4.1.4) using either a global or a locally varying threshold value. We denote binary pixels with values 1 and 0 as *foreground* and *background* pixels, respectively. In most of the following examples, the foreground pixels are shown black and background pixels are shown white, as is common in printing.

At the end of this chapter, we will see that morphological filters are applicable not only to binary images but also to grayscale and even color images, though these operations differ significantly from their binary counterparts.

7.1 Shrink and Let Grow

Our starting point was the observation that a simple 3×3 pixel median filter can round off larger image structures and remove smaller structures, such as points and thin lines, in a binary image. This could for one be useful to eliminate structures that are below a certain size (e. g., to clean an image from noise or dirt). But how can we control the size and possibly the shape of the structures affected by such an operation?

Although its structural effects may be interesting, we disregard the median filter at this point and start with this task again from the beginning. Let's assume that we want to remove small structures from a binary image without significantly altering the remaining larger structures. The key idea for accomplishing this could be the following (Fig. 7.2):

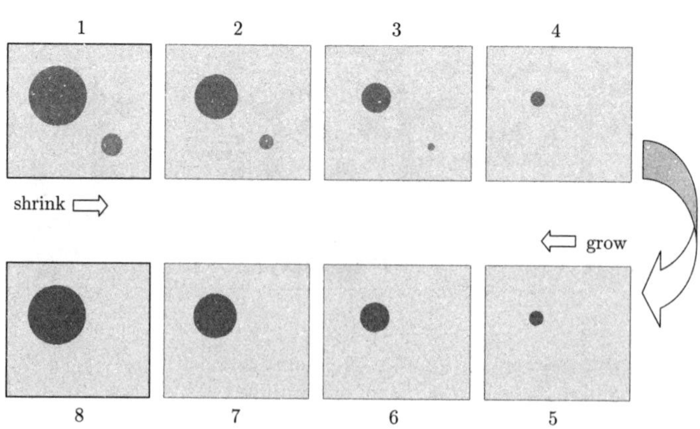

Figure 7.2 Removing small image structures by stepwise shrinking and subsequent growing.

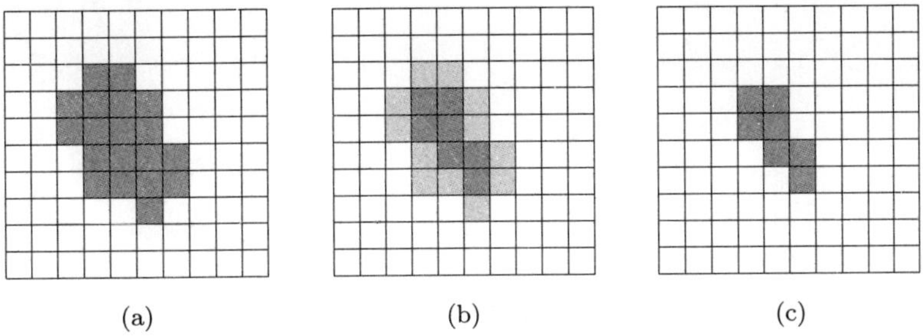

Figure 7.3 "Shrinking" a foreground region by removing a layer of border pixels: original image (a), identified foreground pixels that are in direct contact with the background (b), and result after shrinking (c).

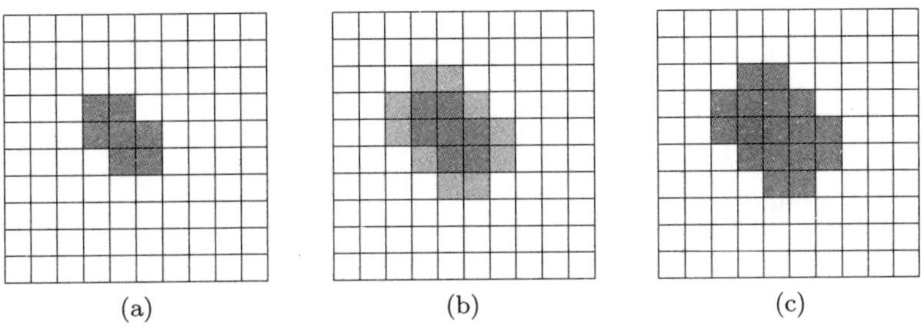

Figure 7.4 "Growing" a foreground region by attaching a layer of pixels: original image (a), identified background pixels that are in direct contact with the region (b), and result after growing (c).

1. First, all structures in the image are iteratively "shrunk" by peeling off a layer of a certain thickness around the boundaries.
2. Shrinking removes the smaller structures step by step, and only the larger structures remain.
3. The remaining structures are then grown back by the same amount.
4. Eventually the larger regions should have returned to approximately their original shapes, while the smaller regions have disappeared from the image.

All we need for this are two types of operations. "Shrinking" means to remove a layer of pixels from a foreground region around all its borders against the background (Fig. 7.3). The other way around, "growing", adds a layer of pixels around the border of a foreground region (Fig. 7.4).

7.1.1 Neighborhood of Pixels

For both operations, we must define the meaning of two pixels being adjacent (i. e., being "neighbors"). Two definitions of "neighborhood" are commonly used

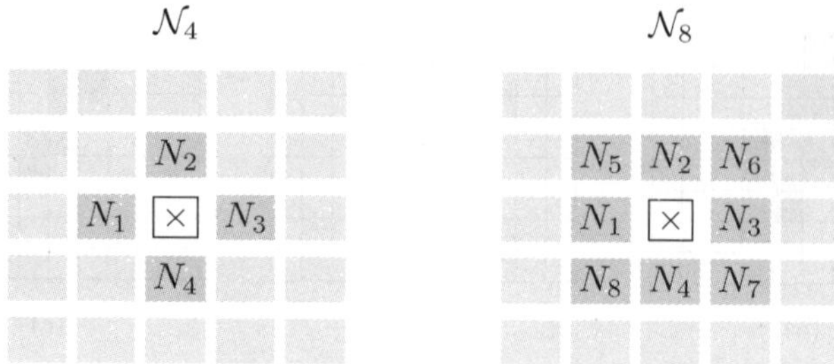

Figure 7.5 Definitions of "neighborhood" on a rectangular pixel grid: *4-neighborhood* $\mathcal{N}_4 = \{N_1, \ldots N_4\}$ (left) and *8-neighborhood* $\mathcal{N}_8 = \mathcal{N}_4 \cup \{N_5, \ldots N_8\}$ (right).

for rectangular pixel grids (Fig. 7.5):

- **4-neighborhood** ($\mathcal{N}_4$): the four pixels adjacent to a given pixel in the horizontal and vertical directions;
- **8-neighborhood** ($\mathcal{N}_8$): the pixels contained in $\mathcal{N}_4$ plus the four adjacent pixels along the diagonals.

7.2 Basic Morphological Operations

Shrinking and growing are indeed the two most basic morphological operations, which are referred to as "erosion" and "dilation", respectively. These morphological operations, however, are much more general than illustrated in the example above. They go well beyond removing or attaching single pixel layers and—in combination—can perform much more complex operations.

7.2.1 The Structuring Element

Similar to the coefficient matrix of a linear filter (see Sec. 5.2), the properties of a morphological filter are specified by elements in a matrix called a "structuring element". In binary morphology, the structuring element (just like the image itself) contains only the values 0 and 1,

$$H(i,j) \in \{0,1\},$$

and the *hot spot* marks the origin of the coordinate system of H (Fig. 7.6). Notice that the hot spot is not necessarily located at the center of the structuring element, nor must its value be 1.

Figure 7.6 Binary structuring element (example). 1–elements are marked with •; 0–cells are empty.

$I \equiv \mathcal{Q}_I = \{(1,1),(2,1),(2,2)\}$ $\quad$ $H \equiv \mathcal{Q}_H = \{(0,0),(1,0)\}$

Figure 7.7 A binary image I or a structuring element H can each be described as a set of coordinate pairs, $\mathcal{Q}_I$ and $\mathcal{Q}_H$, respectively. The dark shaded element in H marks the coordinate origin (hot spot).

7.2.2 Point Sets

For the formal specification of morphological operations, it is helpful to describe binary images as *sets* of two-dimensional coordinate points. For a binary image $I(u,v) \in \{0,1\}$, the corresponding point set $\mathcal{Q}_I$ consists of the coordinate pairs $\boldsymbol{p} = (u,v)$ of all foreground pixels,

$$\mathcal{Q}_I = \{\boldsymbol{p} \mid I(\boldsymbol{p}) = 1\}. \tag{7.1}$$

Of course, as shown in Fig. 7.7, not only a binary image I but also a structuring element H can be described as a point set.

Given a description as point sets, fundamental operations on binary images can also be expressed as simple set operations. For example, *inverting* a binary image $I \to \bar{I}$ (i.e., exchanging foreground and background) is equivalent to building the *complementary* set

$$\mathcal{Q}_{\bar{I}} = \bar{\mathcal{Q}}_I = \{\boldsymbol{p} \in \mathbb{Z}^2 \mid \boldsymbol{p} \notin \mathcal{Q}_I\}. \tag{7.2}$$

Combining two binary images I_1 and I_2 by an OR operation between corresponding pixels, the resulting point set is the *union* of the individual point sets $\mathcal{Q}_{I_1}$ and $\mathcal{Q}_{I_2}$; that is,

$$\mathcal{Q}_{I_1 \vee I_2} = \mathcal{Q}_{I_1} \cup \mathcal{Q}_{I_2}. \tag{7.3}$$

Since a point set $\mathcal{Q}_I$ is only an alternative representation of the binary image I (i.e., $I \equiv \mathcal{Q}_I$), we will use both image and set notations synonymously in the following. For example, we simply write $\bar{I}$ instead of $\bar{\mathcal{Q}}_I$ for an inverted image as in Eqn. (7.2) or $I_1 \cup I_2$ instead of $\mathcal{Q}_{I_1} \cup \mathcal{Q}_{I_2}$ in Eqn. (7.3). The meaning should always be clear in the given context.

Translating (shifting) the binary image I by some coordinate vector $\boldsymbol{d}$ creates a new image with the content $I_{\boldsymbol{d}}(\boldsymbol{p}+\boldsymbol{d}) = I(\boldsymbol{p})$, which corresponds to all coordinates in the point set $\mathcal{Q}_I$ being shifted by $\boldsymbol{d}$; i. e.,

$$I_{\boldsymbol{d}} \equiv \big\{(\boldsymbol{p}+\boldsymbol{d}) \mid \boldsymbol{p} \in I\big\}. \tag{7.4}$$

In some cases, it is also necessary to *reflect* (mirror) a binary image or point set about its origin, which we denote as

$$H^{*} \equiv \{-\boldsymbol{p} \mid \boldsymbol{p} \in H\}. \tag{7.5}$$

7.2.3 Dilation

A *dilation* is the morphological operation that corresponds to our intuitive concept of "growing" as discussed above. As a set operation, it is defined as

$$I \oplus H \equiv \big\{(\boldsymbol{p}+\boldsymbol{q}) \mid \text{for every } \boldsymbol{p} \in I, \boldsymbol{q} \in H\big\}. \tag{7.6}$$

Thus the point set produced by a dilation is the (vector) sum of all possible pairs of coordinate points from the original sets I and H, as illustrated by a simple example in Fig. 7.8.

Alternatively, one could view the dilation as the structuring element H being *replicated* at each foreground pixel of the image I or, conversely, the image I being replicated at each foreground element of H. Expressed in set notation,[1] this is

$$I \oplus H \equiv \bigcup_{\boldsymbol{p} \in I} H_{\boldsymbol{p}} = \bigcup_{\boldsymbol{q} \in H} I_{\boldsymbol{q}}, \tag{7.7}$$

with $H_{\boldsymbol{p}}, I_{\boldsymbol{q}}$ denoting the sets H, I shifted by $\boldsymbol{p}$ and $\boldsymbol{q}$, respectively (see Eqn. (7.4)).

7.2.4 Erosion

The quasi-inverse of dilation is the *erosion* operation, again defined in set notation as

$$I \ominus H \equiv \big\{\boldsymbol{p} \in \mathbb{Z}^2 \mid (\boldsymbol{p}+\boldsymbol{q}) \in I, \text{ for every } \boldsymbol{q} \in H\big\}. \tag{7.8}$$

This definition may appear quite cryptic but is simply explained as follows. A position $\boldsymbol{p}$ is contained in the result $I \ominus H$ if (and only if) the structuring element H—when placed at this position $\boldsymbol{p}$—is *fully contained* in the foreground pixels of the original image; i. e., if $H_{\boldsymbol{p}}$ is a subset of I. Equivalent to Eqn. (7.8), we could thus define binary erosion as

$$I \ominus H \equiv \{\boldsymbol{p} \in \mathbb{Z}^2 \mid H_{\boldsymbol{p}} \subseteq I\}. \tag{7.9}$$

Figure 7.9 shows a simple example for binary erosion.

---

[1] Also see Sec. A.2.

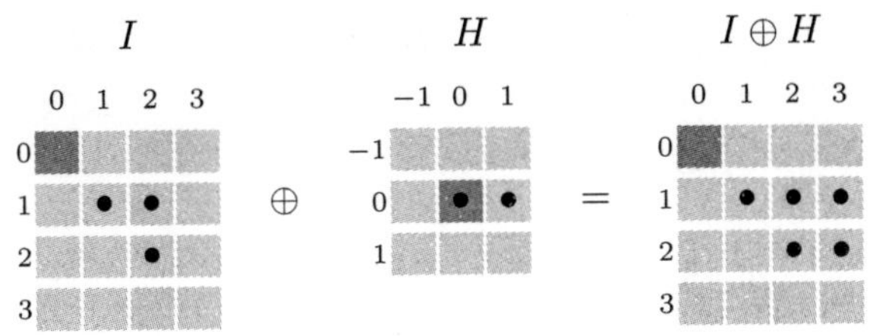

$$I \equiv \{(1,1),(2,1),(2,2)\}, \; H \equiv \{(\mathbf{0},\mathbf{0}),(\mathbf{1},\mathbf{0})\}$$

$$\begin{aligned} I \oplus H \equiv \{ & (1,1)+(\mathbf{0},\mathbf{0}),\ (1,1)+(\mathbf{1},\mathbf{0}), \\ & (2,1)+(\mathbf{0},\mathbf{0}),\ (2,1)+(\mathbf{1},\mathbf{0}), \\ & (2,2)+(\mathbf{0},\mathbf{0}),\ (2,2)+(\mathbf{1},\mathbf{0}) \} \end{aligned}$$

Figure 7.8 Dilation example. The binary image I is subject to dilation with the structuring element H. In the result $I \oplus H$, the structuring element H is replicated at every foreground pixel of the original image I.

7.2.5 Properties of Dilation and Erosion

The dilation operation is *commutative*,

$$I \oplus H = H \oplus I\,, \tag{7.10}$$

and therefore—just as in linear convolution—the image and the structuring element (filter) can be exchanged to get the same result. Dilation is also *associative*,

$$(I_1 \oplus I_2) \oplus I_3 = I_1 \oplus (I_2 \oplus I_3), \tag{7.11}$$

and therefore the ordering of multiple dilations is not relevant. This also means—analogous to linear filters (cf. Eqn. (5.21))—that a dilation with a large structuring element of the form $H_{\mathrm{big}} = H_1 \oplus H_2 \oplus \ldots \oplus H_K$ can be efficiently implemented as a sequence of multiple dilations with smaller structuring elements by

$$I \oplus H_{\mathrm{big}} = (\ldots((I \oplus H_1) \oplus H_2) \oplus \ldots \oplus H_K). \tag{7.12}$$

There is also a *neutral element* δ for the dilation operation, similar to the Dirac function for the linear convolution (see Sec. 5.3.4),

$$I \oplus \delta = \delta \oplus I = I, \qquad \text{with } \delta \equiv \big\{(0,0)\big\}. \tag{7.13}$$

The *erosion* operation is, in contrast to dilation (but similar to arithmetic subtraction), *not* commutative; i.e.,

$$I \ominus H \neq H \ominus I \tag{7.14}$$

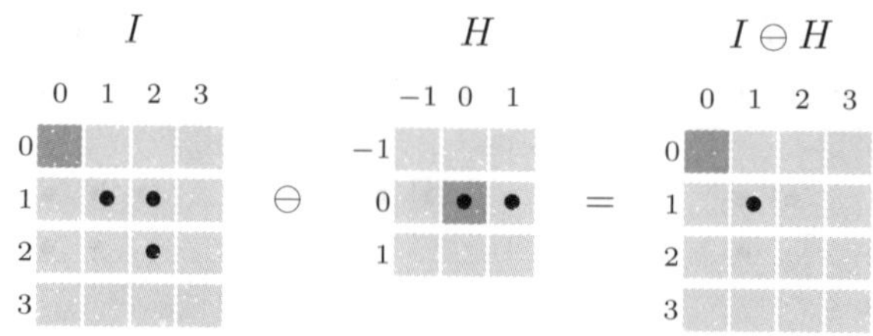

$$I \equiv \{(1,1),(2,1),(2,2)\},\ H \equiv \{(\mathbf{0},\mathbf{0}),(\mathbf{1},\mathbf{0})\}$$

$$I \ominus H \equiv \{\,(1,1)\,\} \text{ because}$$

$$(1,1)+(\mathbf{0},\mathbf{0}) = (1,1) \in I \quad \textbf{and} \quad (1,1)+(\mathbf{1},\mathbf{0}) = (2,1) \in I$$

Figure 7.9 Erosion example. The binary image I is subject to erosion with H as the structuring element. H is only covered by I when placed at position $\boldsymbol{p} = (1,1)$. Thus the resulting point set contains only the single coordinate $(1,1)$.

in general. However, if erosion and dilation are combined, then—again in analogy with arithmetic subtraction and addition—the following chain rule holds:

$$(I_1 \ominus I_2) \ominus I_3 \;=\; I_1 \ominus (I_2 \oplus I_3). \tag{7.15}$$

Although dilation and erosion are not mutually inverse (in general, the effects of dilation cannot be undone by a subsequent erosion), there are still some strong formal relations between these two operations.

For one, dilation and erosion are *dual* in the sense that a dilation of the *foreground* (I) can be accomplished by an erosion of the *background* ($\bar{I}$) and subsequent inversion of the result,

$$I \oplus H \;\equiv\; \overline{(\bar{I} \ominus H^*)}, \tag{7.16}$$

where H^* denotes the *reflection* of H (Eqn. (7.5)). This works similarly the other way, too, namely

$$I \ominus H \;\equiv\; \overline{(\bar{I} \oplus H^*)}, \tag{7.17}$$

effectively eroding the foreground by dilating the background with the mirrored structuring element, as illustrated by the example in Fig. 7.10 (see [17, pp. 521–524] for a proof).

Equation (7.17) is interesting because it shows that we only need to implement either dilation or erosion for computing both, considering that the foreground-background inversion is a very simple task. Algorithm 7.1 gives a simple algorithmic description of dilation and erosion based on the relationships above. The corresponding Java implementation is shown later, in Sec. 7.5.2.

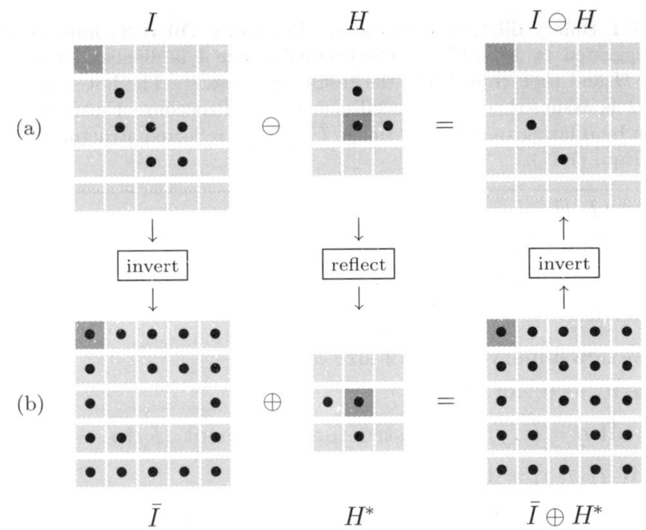

Figure 7.10 Implementing erosion via dilation. The binary erosion of the foreground $I \ominus H$ (a) can be implemented by dilating the inverted (background) image $\bar{I}$ with the reflected structuring element H^* and subsequently inverting the result again (b).

7.2.6 Designing Morphological Filters

A morphological filter is unambiguously specified by (a) the type of operation and (b) the contents of the structuring element. The appropriate size and shape of the structuring element depends upon the application, image resolution, etc. In practice, structuring elements of quasi-circular shape are frequently used, such as the examples shown in Fig. 7.11.

A dilation with a circular (disk-shaped) structuring element with radius r adds a layer of thickness r to any foreground structure in the image. Conversely, an erosion with that structuring element peels off layers of the same thickness. Figure 7.13 shows the results of dilation and erosion with disk-shaped structur-

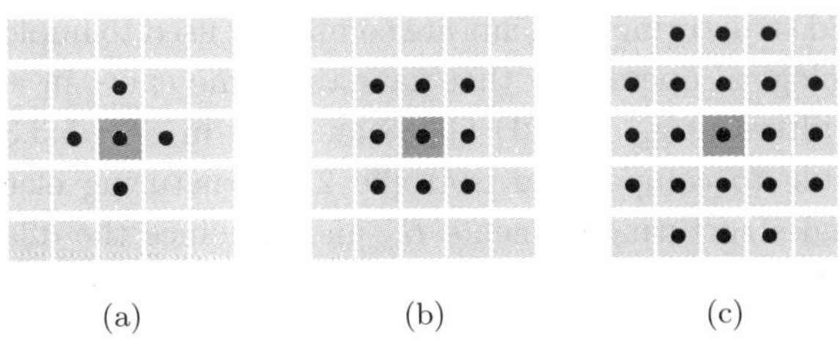

Figure 7.11 Typical small structuring elements: 4-neighborhood (a), 8-neighborhood (b), and "small disk" (c).

Algorithm 7.1 Binary dilation and erosion. Procedure DILATE() implements the binary dilation as suggested by Eqn. (7.7). The original image I is displaced to each foreground coordinate of H and then copied into the resulting image I'. The hot spot of the structuring element H is assumed to be at coordinate $(0, 0)$. Procedure ERODE() implements the binary erosion by dilating the inverted image $\bar{I}$ with the reflected structuring element H^*, as described by Eqn. (7.17).

DILATE (I, H)
  I: binary image
  H: binary structuring element
  Returns the dilated image $I' = I \oplus H$
 $I' \leftarrow$ new binary image of size $w \times h$
 $I'(u,v) \leftarrow 0$, for all (u,v) ▷ $I' \leftarrow \varnothing$
 for all $\boldsymbol{q} = (i,j)$ in the structuring element H **do**
  if $H(i,j) = 1$ **then**   ▷ $\boldsymbol{q} \in H$
   MERGE THE SHIFTED IMAGE $I_{\boldsymbol{q}}$ WITH I':   ▷ $I' \leftarrow I' \cup I_{\boldsymbol{q}}$
   for all image locations $\boldsymbol{p} = (u,v)$ **do**
    if $I(u,v) = 1$ **then**   ▷ $\boldsymbol{p} \in I$
     $I'(u{+}i, v{+}j) \leftarrow 1$ ▷ $I' \leftarrow I' \cup \{(\boldsymbol{p}{+}\boldsymbol{q})\}$
 return I'.

ERODE (I, H)
  I: binary image
  H: binary structuring element
  Returns the eroded image $I' = I \ominus H$
 $\bar{I} \leftarrow$ INVERT(I) ▷ $\bar{I} \leftarrow \neg I$
 $H^* \leftarrow$ REFLECT(H)
 return INVERT(DILATE$(\bar{I}, H^*)$).   ▷ $I \ominus H = \overline{(\bar{I} \oplus H^*)}$

ing elements of different diameters applied to the original image in Fig. 7.12. Dilation and erosion results for various other structuring elements are shown in Fig. 7.14.

Disk-shaped structuring elements are commonly used to implement *isotropic* filters, morphological operations that have the same effect in every direction. Unlike linear filters (e. g., the 2D Gaussian filter in Sec. 5.3.3), it is generally not possible to compose an isotropic 2D structuring element H° from one-dimensional structuring elements H_x and H_y since the dilation $H_x \oplus H_y$ always results in a rectangular (i. e., nonisotropic) structure. A remedy for approximating large disk-shaped filters is to alternately apply smaller disk-shaped operators of different shapes, as illustrated in Fig. 7.15. The resulting filter is generally not fully isotropic but can be implemented efficiently as a sequence

Figure 7.12 Original binary image and the section used in the following examples (illustration by Albrecht Dürer, 1515).

of small filters.

7.2.7 Application Example: Outline

A typical application of morphological operations is to extract the boundary pixels of the foreground structures. The process is very simple. First, we apply an erosion on the original image I to remove the boundary pixels of the foreground,

$$I' = I \ominus H_n,$$

using the 4- or 8-neighborhood (Fig. 7.11) as the structuring element H_n. To extract the actual boundary pixels B, we take the intersection of the original image I and the inverted result $\bar{I}'$; that is,

$$B = I \cap \overline{I'} = I \cap \overline{(I \ominus H_n)}. \qquad (7.18)$$

Notice that using the 4-neighborhood as the structuring element H_n produces "8-connected" contours and vice versa [23, p. 504].

The process of boundary extraction is illustrated on a simple example in Fig. 7.16. As can be observed in this figure, the result B contains exactly those pixels that are *different* in the original image I and the eroded image $I' = I \ominus H_n$, which can also be obtained by an exclusive-OR (XOR) operation between pairs of pixels; that is, boundary extraction from a binary image can be implemented as

$$B(u,v) = \mathrm{XOR}\big(I(u,v), I'(u,v)\big) \quad \text{for all } (u,v). \qquad (7.19)$$

Figure 7.17 shows a more complex example for isolating the boundary pixels in a real image.

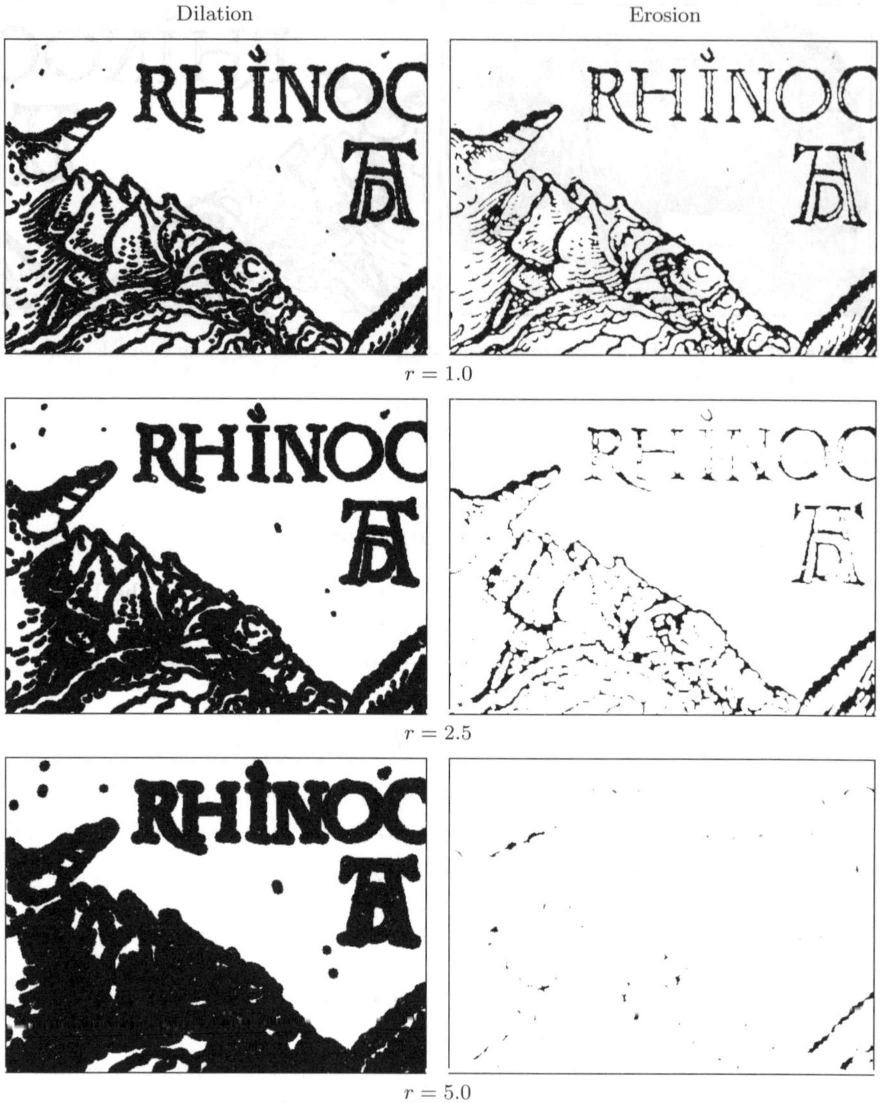

Figure 7.13 Results of binary dilation and erosion with disk-shaped structuring elements. The radius of the disk (r) is 1.0 (top), 2.5 (center), or 5.0 (bottom).

7.3 Composite Operations

Due to their semiduality, dilation and erosion are often used together in composite operations, two of which are so important that they even carry their own names and symbols: "opening" and "closing". They are probably the most frequently used morphological operations in practice.

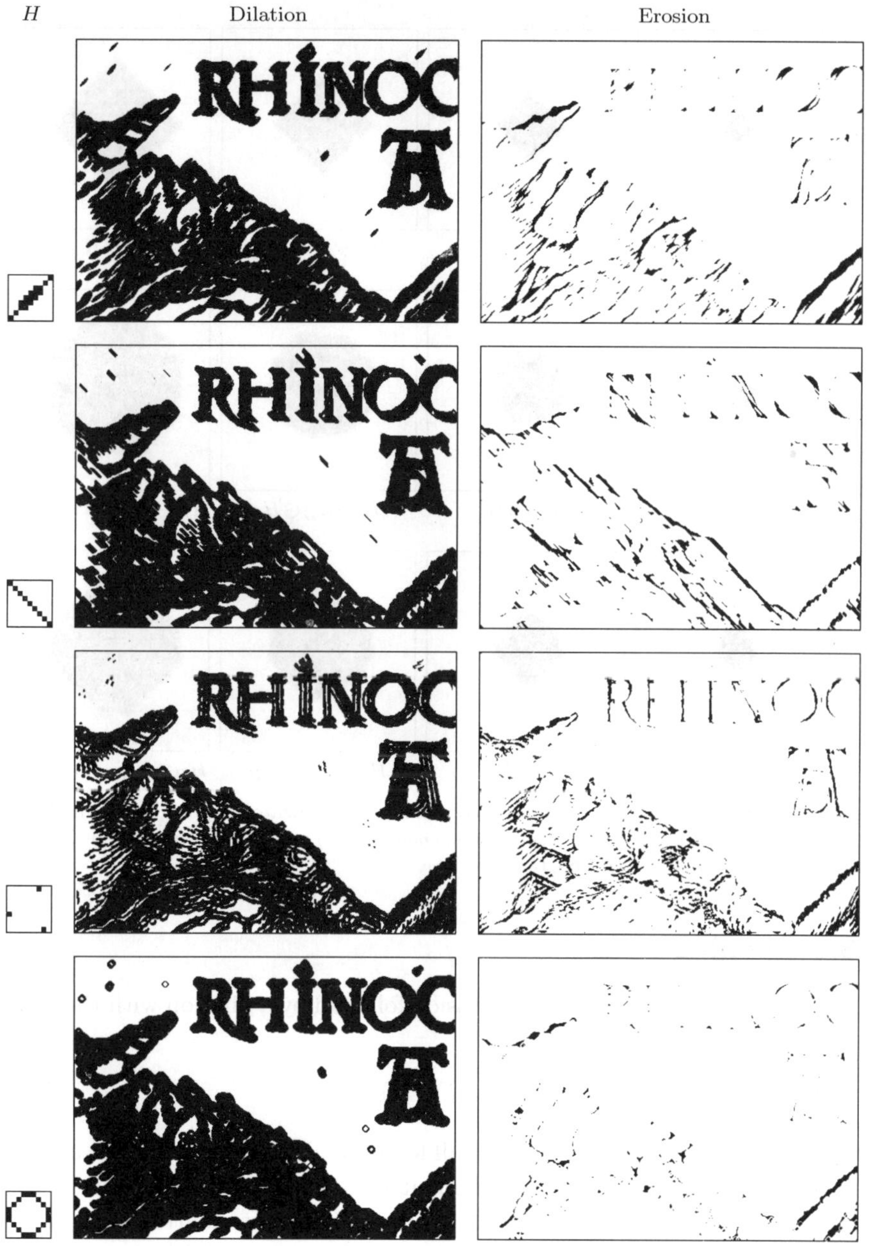

Figure 7.14 Examples of binary dilation and erosion with various free-form structuring elements. The structuring elements H are shown in the left column (enlarged). Notice that the dilation expands every isolated foreground point to the shape of the structuring element, analogous to the *impulse response* of a linear filter. Under erosion, only those elements where the structuring element is fully contained in the original image survive.

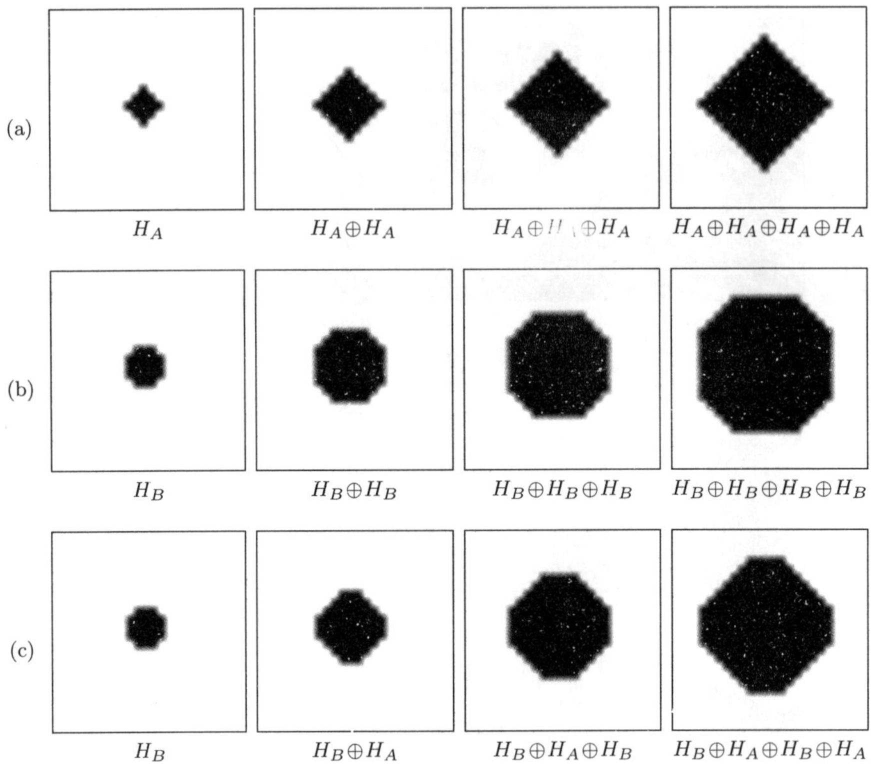

Figure 7.15 Composition of large morphological filters by repeated application of smaller filters: repeated application of the structuring element H_A (a) and structuring element H_B (b); alternating application of H_B and H_A (c).

7.3.1 Opening

A binary opening $I \circ H$ denotes an erosion followed by a dilation with the *same* structuring element H,

$$I \circ H = (I \ominus H) \oplus H. \tag{7.20}$$

The main effect of an opening is that all foreground structures that are smaller than the structuring element are eliminated in the first step (erosion). The remaining structures are smoothed by the subsequent dilation and grown back to approximately their original size, as demonstrated by the examples in Fig. 7.18). This process of shrinking and subsequent growing corresponds to the idea for eliminating small structures that we had initially sketched in Sec. 7.1.

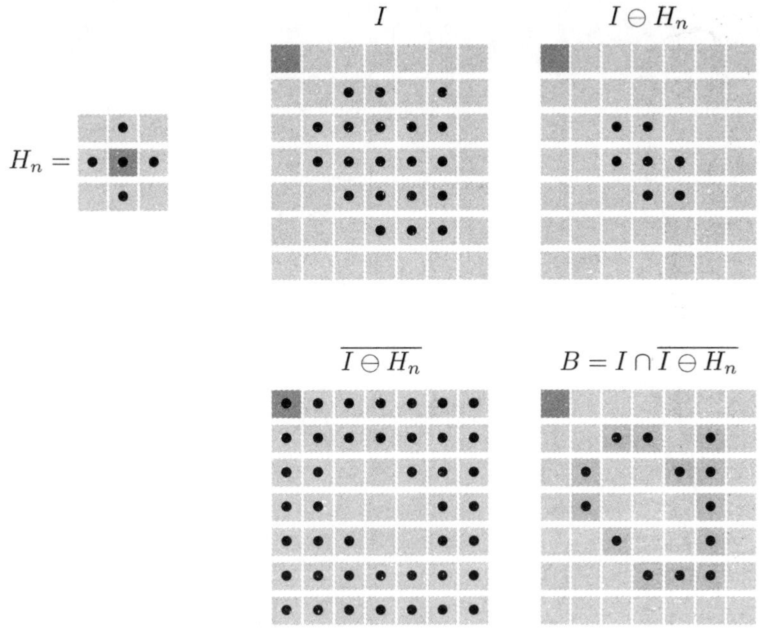

Figure 7.16 Outline example using a 4-neighborhood structuring element H_n. The image I is first eroded $(I \ominus H_n)$ and subsequently inverted $(\overline{I \ominus H_n})$. The boundary pixels are finally obtained as the intersection $I \cap \overline{I \ominus H_n}$.

7.3.2 Closing

When the sequence of erosion and dilation is reversed, the resulting operation is called a closing and denoted $I \bullet H$,

$$I \bullet H = (I \oplus H) \ominus H. \tag{7.21}$$

A *closing* removes (closes) holes and fissures in the foreground structures that are smaller than the structuring element H. Some examples with typical disk-shaped structuring elements are shown in Fig. 7.18.

7.3.3 Properties of Opening and Closing

Both operations, opening as well as closing, are *idempotent*, meaning that their results are "final" in the sense that any subsequent application of the same operation no longer changes the result; i. e.,

$$\begin{aligned} I \circ H &= (I \circ H) \circ H = ((I \circ H) \circ H) \circ H = \ldots , \\ I \bullet H &= (I \bullet H) \bullet H = ((I \bullet H) \bullet H) \bullet H = \ldots . \end{aligned} \tag{7.22}$$

(a) (b)

Figure 7.17 Extraction of boundary pixels using morphological operations. The 4-neighborhood structuring element used in (a) produces 8-connected contours. Conversely, using the 8-neighborhood as the structuring element gives 4-connected contours (b).

Also, opening and closing are "duals" in the sense that opening the foreground is equivalent to closing the background and vice versa; i. e.,

$$I \circ H = \overline{(\bar{I} \bullet H)} \quad \text{and} \quad I \bullet H = \overline{(\bar{I} \circ H)}. \tag{7.23}$$

7.4 Grayscale Morphology

Morphological operations are not confined to binary images but are also for intensity (grayscale) images. In fact, the definition of grayscale morphology is a *generalization* of binary morphology, with the binary OR and AND operators replaced by the arithmetic MAX and MIN operators, respectively. As a consequence, procedures designed for grayscale morphology can also perform binary

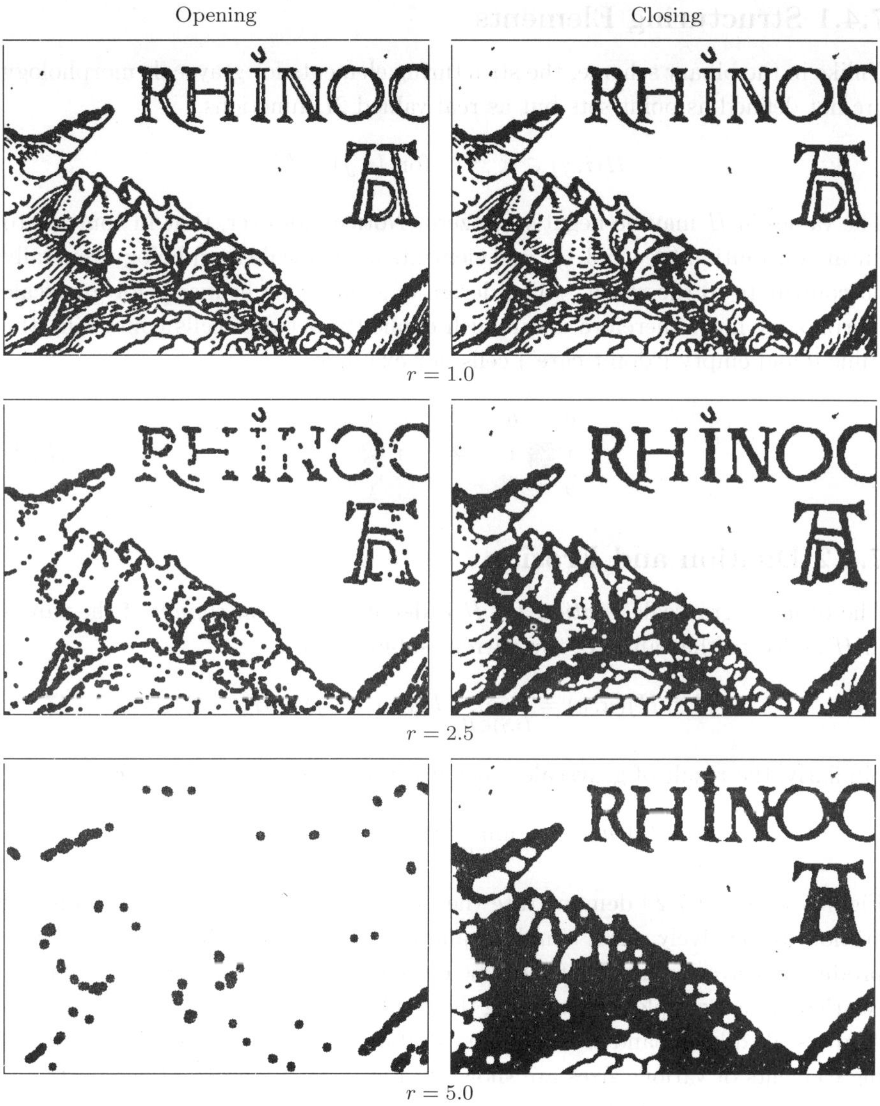

Figure 7.18 Binary opening and closing with disk-shaped structuring elements. The radius r of the structuring element H is 1.0 (top), 2.5 (center), or 5.0 (bottom).

morphology (but not the other way around).[2] In the case of color images, the grayscale operations are usually applied individually to each color channel.

[2] ImageJ provides a single implementation of morphological operations that handles both binary and grayscale images (see Sec. 7.5.5).

7.4.1 Structuring Elements

Unlike in the binary scheme, the structuring elements for grayscale morphology are not defined as point sets but as real-valued 2D functions,

$$H(i,j) \in \mathbb{R}\,, \qquad \text{for } \ (i,j) \in \mathbb{Z}^2.$$

The values in H may be negative or zero. Notice, however, that in contrast to linear convolution (Sec. 5.3.1), zero elements in grayscale morphology generally *do* contribute to the result.[3] The design of structuring elements for grayscale morphology must therefore distinguish explicitly between cells containing the value 0 and empty ("don't care") cells; for example

$$\begin{array}{|c|c|c|}\hline 0 & 1 & 0 \\ \hline 1 & 2 & 1 \\ \hline 0 & 1 & 0 \\ \hline \end{array} \;\neq\; \begin{array}{|c|c|c|}\hline & 1 & \\ \hline 1 & 2 & 1 \\ \hline & 1 & \\ \hline \end{array}\,. \tag{7.24}$$

7.4.2 Dilation and Erosion

The result of grayscale *dilation* $I \oplus H$ is defined as the *maximum* of the values in H added to the values of the current subimage of I,

$$(I \oplus H)(u,v) = \max_{(i,j)\in H} \bigl\{ I(u{+}i, v{+}j) + H(i,j) \bigr\}. \tag{7.25}$$

Similarly, the result of grayscale *erosion* is the minimum of the differences,

$$(I \ominus H)(u,v) = \min_{(i,j)\in H} \bigl\{ I(u{+}i, v{+}j) - H(i,j) \bigr\}. \tag{7.26}$$

Figures 7.19 and 7.20 demonstrate the basic process of grayscale dilation and erosion, respectively, on a simple example. In general, either operation may produce *negative* results that must be considered if the range of pixel values is restricted; for example, by clamping the results (see Sec. 4.1.2). Some examples of grayscale dilation and erosion on natural images using disk-shaped structuring elements of various sizes are shown in Fig. 7.21. Figure 7.22 demonstrates the same operations with some freely designed structuring elements.

7.4.3 Grayscale Opening and Closing

Opening and closing on grayscale images are defined, identical to the binary case (Eqns. (7.20) and (7.21)), as operations composed of dilation and erosion with the same structuring element. Some examples are shown in Fig. 7.23 for disk-shaped structuring elements and in Fig. 7.24 for various nonstandard

[3] While a zero coefficient in a linear convolution matrix simply means that the corresponding image pixel is ignored.

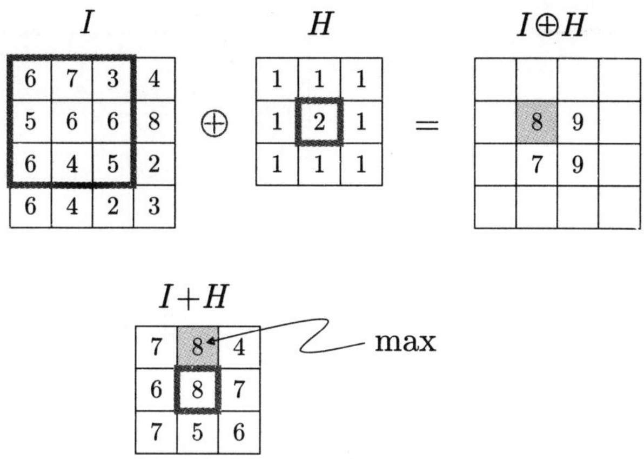

Figure 7.19 Grayscale dilation $I \oplus H$. The 3×3 pixel structuring element H is placed on the image I in the upper left position. Each value of H is added to the corresponding element of I; the intermediate result $(I + H)$ for this particular position is shown below. Its maximum value $8 = 7 + 1$ is inserted into the result $(I \oplus H)$ at the current position of the filter origin. The results for three other filter positions are also shown.

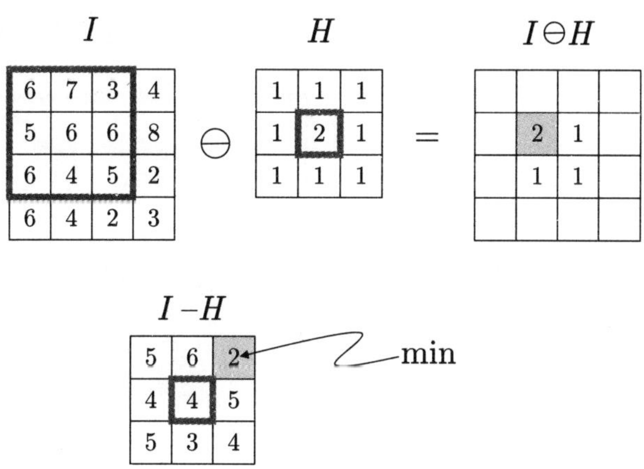

Figure 7.20 Grayscale erosion $I \ominus H$. The 3×3 pixel structuring element H is placed on the image I in the upper left position. Each value of H is subtracted from the corresponding element of I; the intermediate result $(I - H)$ for this particular position is shown below. Its minimum value $3 - 1 = 2$ is inserted into the result $(I \ominus H)$ at the current position of the filter origin. The results for three other filter positions are also shown.

structuring elements. Notice that interesting effects can be obtained, particularly from structuring elements resembling the shape of brush or other stroke patterns.

Figure 7.21 Grayscale dilation and erosion with disk-shaped structuring elements. The radius r of the structuring element is 2.5 (top), 5.0 (center), or 10.0 (bottom).

7.5 Implementing Morphological Filters

7.5.1 Binary Images in ImageJ

In ImageJ, binary images contain 8 bits per pixel, the same as ordinary grayscale images.[4] A zero intensity value is interpreted as a binary 0, and

[4] ImageJ does not provide a special (1-bit) data format for binary images. The class `BinaryProcessor` keeps image data as byte (8-bit) arrays, as does `ByteProcessor`

Figure 7.22 Grayscale dilation and erosion with various free-form structuring elements.

any value greater than zero is considered a binary 1. Usually the intensity values 0 and 255 are used to represent the binaries 0 and 1, respectively, in which case the background pixels are displayed black and the foreground pixels

for grayscale images.

Figure 7.23 Grayscale opening and closing with disk-shaped structuring elements. The radius r of the structuring element is 2.5 (top), 5.0 (center), or 10.0 (bottom).

are white by default. If an inverted display (black foreground) is desired, this can be easily accomplished by inverting the display function or lookup table (LUT) either interactively through the menu

Image→Lookup Tables→Invert LUT

or within the Java program by invoking the `ImageProcessor` method

```
void invertLut()
```

Figure 7.24 Grayscale opening and closing with various free-form structuring elements.

on the corresponding `ImageProcessor` object. Any of these instructions changes only the screen presentation of the current image but not its contents (pixel values).

7.5.2 Dilation and Erosion

Most morphological operations are already implemented in ImageJ as methods of the class `ImageProcessor` (see also Sec. 7.5.5); however, they are restricted to structuring elements of size 3×3 pixels.

In the following, we describe a sample implementation of binary *dilation* for arbitrary structuring elements that can be used (due to the duality of dilation and erosion; see Eqn. (7.16)) for implementing most other morphological operations. Input to `dilate()` is a binary image `I` with values 0 for the background and 255 for the foreground[5] and a two-dimensional structuring element `H` with 0/1-values whose origin (hot spot) is assumed at its center:

```
1  import ij.process.Blitter;
2  import ij.process.ImageProcessor;
3  ...
4  void dilate(ImageProcessor I, int[][] H){
5    //assume that the hot spot of H is at its center (ic,jc):
6    int ic = (H[0].length-1)/2;
7    int jc = (H.length-1)/2;
8
9    //create a temporary (empty) image:
10   ImageProcessor tmp
11           = I.createProcessor(I.getWidth(),I.getHeight());
12
13   for (int j=0; j<H.length; j++){
14     for (int i=0; i<H[j].length; i++){
15       if (H[j][i] > 0) { // this pixel is set
16         //copy image into position (i-ic,j-jc):
17         tmp.copyBits(I,i-ic,j-jc,Blitter.MAX);
18       }
19     }
20   }
21   //copy the temporary result back to original image
22   I.copyBits(np,0,0,Blitter.COPY);
23 }
```

The `dilate()` method destructively modifies the input image `I`. First (in line 10), a temporary (empty) image `tmp` of the same size as `I` is created, which is then modified and eventually (line 22) copied back to replace the input image. The actual dilation is performed iteratively by copying a shifted version of the original image into the temporary image `tmp` for every position (i, j) of the structuring element with $H(i, j) > 0$. This is done in line 17 using the `ImageProcessor` method `copyBits()` with `Blitter.MAX` as the operation parameter (see also Sec. 4.8.3). If the pixels are interpreted as binary values, the max-operation corresponds to a logical OR operation between the pixels in the intermediate image `tmp` and the shifted input image `I`.

---

[5] In fact, any value greater than 0 is considered a foreground pixel.

Dilation is the only operation that must be implemented in detail since erosion can be performed as a dilation of the background by inverting the image, performing a dilation, and inverting again (see Alg. 7.1):

```
24  void erode(ImageProcessor I, int[][] H) {
25    ip.invert();
26    dilate(ip, reflect(H));
27    ip.invert();
28  }
```

In the above, the method `reflect(H)` (line 26) returns a mirrored copy of the structuring element H and `invert()` (lines 25, 27) is a standard ImageJ method defined by the class `ImageProcessor`.

7.5.3 Opening and Closing

Opening and closing operations are now easy to implement as combinations of dilation and erosion with the same structuring element `H`, as described in Sec. 7.3:

```
29  void open(ImageProcessor I, int[][] H) {
30    erode(I,H);
31    dilate(I,H);
32  }
```

```
33  void close(ImageProcessor I, int[][] H) {
34    dilate(I,H);
35    erode(I,H);
36  }
```

7.5.4 Outline

To implement the *outline* operation for extracting the boundary pixels, as described in Sec. 7.2.7, we use a 3×3 pixel structuring element `H` to represent the 4-neighborhood. First we create a duplicate (`Ie`) of the input image (`I`), which is then subject to erosion with `H` (line 43). The boundary pixels are obtained by computing the difference between the original and the eroded image (using the standard method `copyBits()` with the argument `Blitter.DIFFERENCE`). In binary terms, this is an exclusive-OR (XOR) operation between the pixels in `I` and `Ie`, which implements the set intersection (see Eqn. (7.19)). The differencing operation in line 44 stores its result in `I`, which finally contains the boundary pixels of the foreground structures:

```
void outline(ImageProcessor I) {
  int[][] H = { //4-neighborhood structuring element
    {0,1,0},
    {1,1,1},
    {0,1,0}};
  ImageProcessor Ie = I.duplicate();
  erode(Ie,H);    // I' ← I ⊖ H
  I.copyBits(Ie,0,0,Blitter.DIFFERENCE); // I ← XOR(I, I')
}
```

7.5.5 Morphological Operations in ImageJ

Class `ImageProcessor`

ImageJ defines several methods for basic morphological operations in the class `ImageProcessor`:

```
void dilate()
void erode()
void open()
void close()
```

All these methods apply a 3×3 pixel box-shaped structuring element (see Fig. 7.11 (b)) and perform either binary or grayscale operations, depending upon the image content. The class `ColorProcessor` uses the same methods for RGB images by processing the color channels individually like ordinary grayscale or binary images.

Class `BinaryProcessor`

The class `BinaryProcessor` (a subclass of `ByteProcessor`) offers the specific morphological methods

```
void outline()
void skeletonize()
```

which are only defined for binary images. The method `outline()` implements the extraction of boundary pixels using an 8-neighborhood structuring element, as described in Sec. 7.2.7.

The operation implemented by the method `skeletonize()` is often referred to as "thinning" or "skeletonization", which iteratively erodes structures down to a thickness of 1 pixel without splitting them. This requires a decision based on the current image content within the filter region (typically of size 3×3 pixels) as to whether another erosion should be applied or not. The operation repeats until no more changes can be made to the result (see, e. g., [17, p. 535] or [24, p. 517] for details). The actual implementation in ImageJ is based on

(a) (b)

Figure 7.25 Example of thinning with the `skeletonize()` method: original image and detail (a) and results from thinning (b).

an efficient algorithm by Zhang and Suen [44], and an example of applying the `skeletonize()` method is shown in Fig. 7.25.

The methods `outline()` and `skeletonize()` are only applicable to objects of type `BinaryProcessor`, which can be created from existing `ByteProcessor` objects. This assumes, however, that the original image contains only values of 0 (background) and 255 (foreground). The following example shows the use of `outline()` within the `run()` method of an ImageJ plugin:

```
public void run(ImageProcessor ip) {
    ByteProcessor byteP
        = (ByteProcessor) ip.convertToByte(true); // scale!
    BinaryProcessor binP
        = new BinaryProcessor(byteP);
    binP.outline();
    ...
}
```

Notice that the new `BinaryProcessor` object `binP` does not allocate any new image data but only references the data of the parent image `byteP`. Thus any subsequent modification to `binP` (e. g., by invoking the method `outline()`) is also visible in `byteP`.

Other morphological filters

In addition to the morphological operations implemented in ImageJ itself, there are additional plugins and complete morphological packages available online,[6] including the morphology operators by Gabriel Landini and the Grayscale Morphology package by Dimiter Prodanov, which allows structuring elements to be interactively specified.

7.6 Exercises

Exercise 7.1

Manually compute the results of dilation and erosion for the following image I and the structuring elements H_1 and H_2:

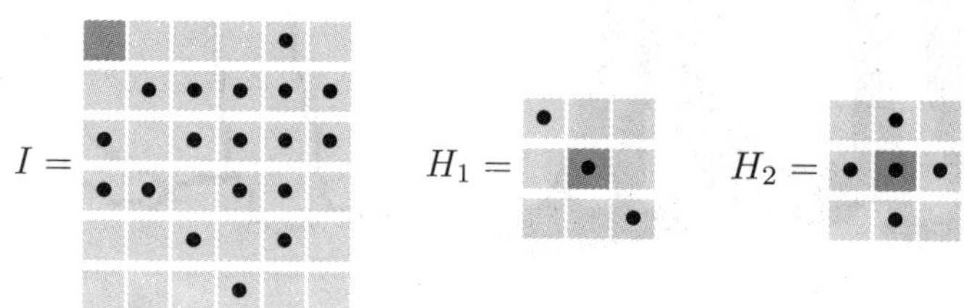

Exercise 7.2

Assume that a binary image I contains unwanted foreground spots with a maximum diameter of 5 pixels that should be removed without damaging the remaining structures. Design a suitable morphological procedure, and evaluate its performance on appropriate test images.

Exercise 7.3

Show that, in the special case of the structuring elements with the contents

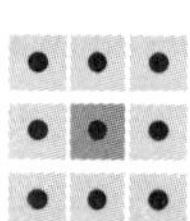 for binary images and 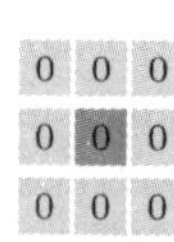

for grayscale images,

dilation is equivalent to a 3×3 pixel maximum filter and erosion is equivalent to a 3 × 3 pixel minimum filter (see Sec. 5.4.1).

---

[6] http://rsb.info.nih.gov/ij/plugins/.

8 Color Images

Color images are involved in every aspect of our lives, where they play an important role in everyday activities such as television, photography, and printing. Color perception is a fascinating and complicated phenomenon that has occupied the interest of scientists, psychologists, philosophers, and artists for hundreds of years [38, 39]. In this chapter, we focus on those technical aspects of color that are most important for working with digital color images. Our emphasis will be on understanding the various representations of color and correctly utilizing them when programming. Additional color-related issues, such as color quantization and colorimetric color spaces, are covered in Volume 2 [6].

8.1 RGB Color Images

The RGB color schema encodes colors as combinations of the three primary colors: red (R), green (G), and blue (B). This scheme is widely used for transmission, representation, and storage of color images on both analog devices such as television sets and digital devices such as computers, digital cameras, and scanners. For this reason, many image-processing and graphics programs use the RGB schema as their internal representation for color images, and most language libraries, including Java's imaging APIs, use it as their standard image representation.

RGB is an *additive* color system, which means that all colors start with black and are created by adding the primary colors. You can think of color formation in this system as occurring in a dark room where you can overlay

W. Burger, M.J. Burge, *Principles of Digital Image Processing*, Undergraduate Topics in Computer Science, DOI 10.1007/978-1-84800-191-6_8, © Springer-Verlag London Limited, 2009

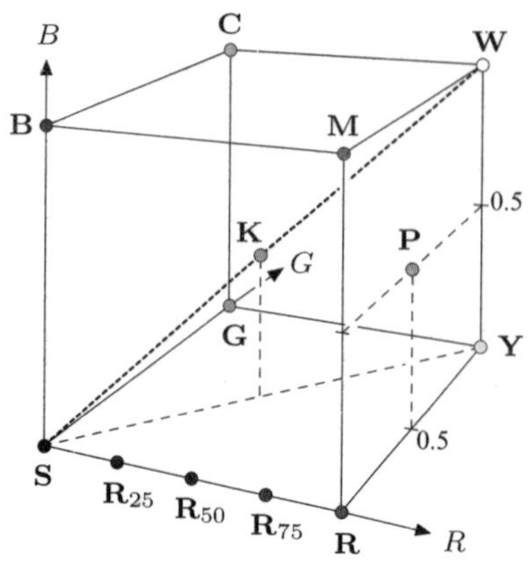

Point	Color	RGB Value R	G	B
S	Black	0.00	0.00	0.00
R	Red	1.00	0.00	0.00
Y	Yellow	1.00	1.00	0.00
G	Green	0.00	1.00	0.00
C	Cyan	0.00	1.00	1.00
B	Blue	0.00	0.00	1.00
M	Magenta	1.00	0.00	1.00
W	White	1.00	1.00	1.00
K	50% Gray	0.50	0.50	0.50
$\mathbf{R}_{75}$	75% Red	0.75	0.00	0.00
$\mathbf{R}_{50}$	50% Red	0.50	0.00	0.00
$\mathbf{R}_{25}$	25% Red	0.25	0.00	0.00
P	Pink	1.00	0.50	0.50

Figure 8.1 Representation of the RGB color space as a three-dimensional unit cube. The primary colors red (R), green (G), and blue (B) form the coordinate system. The "pure" red color (**R**), green (**G**), blue (**B**), cyan (**C**), magenta (**M**), and yellow (**Y**) lie on the vertices of the color cube. All the shades of gray, of which **K** is an example, lie on the diagonal between black **S** and white **W**.

three beams of light—one red, one green, and one blue—on a sheet of white paper. To create different colors, you would modify the intensity of each of these beams independently. The distinct intensity of each primary color beam controls the shade and brightness of the resulting color. The colors gray and white are created by mixing the three primary color beams at the same intensity. A similar operation occurs on the screen of a color television or CRT[1]-based computer monitor, where tiny, close-lying dots of red, green, and blue phosphorous are simultaneously excited by a stream of electrons to distinct energy levels (intensities), creating a seemingly continuous color image.

The RGB color space can be visualized as a three-dimensional unit cube in which the three primary colors form the coordinate axis. The RGB values are positive and lie in the range $[0, C_{\max}]$; for most digital images, $C_{\max} = 255$. Every possible color $\mathbf{C}_i$ corresponds to a point within the RGB color cube of the form

$$\mathbf{C}_i = (R_i, G_i, B_i),$$

where $0 \leq R_i, G_i, B_i \leq C_{\max}$. RGB values are often normalized to the interval $[0, 1]$ so that the resulting color space forms a unit cube (Fig. 8.1). The point $\mathbf{S} = (0, 0, 0)$ corresponds to the color black, $\mathbf{W} = (1, 1, 1)$ corresponds to the color white, and all the points lying on the diagonal between $\mathbf{S}$ and $\mathbf{W}$ are shades of gray created from equal color components $R = G = B$.

Figure 8.2 shows a color test image and its corresponding RGB color components, displayed here as intensity images. We will refer to this image in a

[1] Cathode ray tube.

Figure 8.2 A color image and its corresponding RGB channels. The fruits depicted are mainly yellow and red and therefore have high values in the R and G channels. In these regions, the B content is correspondingly lower (represented here by darker gray values) except for the bright highlights on the apple, where the color changes gradually to white. The tabletop in the foreground is purple and therefore displays correspondingly higher values in its B channel.

number of examples that follow in this chapter.

RGB is a very simple color system, and as demonstrated in Sec. 8.2, a basic knowledge of it is often sufficient for processing color images or transforming them into other color spaces. At this point, we will not be able to determine what color a particular RGB pixel corresponds to in the real world, or even what the primary colors red, green, and blue truly mean in a physical (i. e., colorimetric) sense. Instead, in this volume we will rely on our intuitive understanding of color and address colorimetry and color spaces in detail in

Vol. 2 [6, Sec. 6].

8.1.1 Organization of Color Images

Color images are represented in the same way as grayscale images, by using an array of pixels in which different models are used to order the individual color components. In the next sections we will examine the difference between *true color* images, which utilize colors uniformly selected from the entire color space, and so-called *palleted* or *indexed* images, in which only a select set of distinct colors are used. Deciding which type of image to use depends on the requirements of the application.

True color images

A pixel in a true color image can represent any color in its color space, as long as it falls within the (discrete) range of its individual color components. True color images are appropriate when the image contains many colors with subtle differences, as occurs in digital photography and photo-realistic computer graphics. Next we look at two methods of ordering the color components in true color images: *component ordering* and *packed ordering.*

Component ordering. In *component ordering* (also referred to as *planar ordering*) the color components are laid out in separate arrays of identical dimensions. In this case, the color image

$$I = (I_R, I_G, I_B)$$

can be thought of as a vector of related intensity images I_R, I_G, and I_B (Fig. 8.3), and the RGB component values of the color image I at position (u, v) are obtained by accessing all three intensity images as follows:

$$\begin{pmatrix} R_{u,v} \\ G_{u,v} \\ B_{u,v} \end{pmatrix} \leftarrow \begin{pmatrix} I_R(u,v) \\ I_G(u,v) \\ I_B(u,v) \end{pmatrix}. \tag{8.1}$$

Packed ordering. In *packed ordering*, the component values that represent the color of a particular pixel are packed together into a single element of the image array (Fig. 8.4) so that

$$I(u, v) = (R_{u,v}, G_{u,v}, B_{u,v}).$$

The RGB value of a packed image I at the location (u, v) is obtained by accessing the individual components of the color pixel as

$$\begin{pmatrix} R_{u,v} \\ G_{u,v} \\ B_{u,v} \end{pmatrix} \leftarrow \begin{pmatrix} \mathsf{Red}(I(u,v)) \\ \mathsf{Green}(I(u,v)) \\ \mathsf{Blue}(I(u,v)) \end{pmatrix}. \tag{8.2}$$

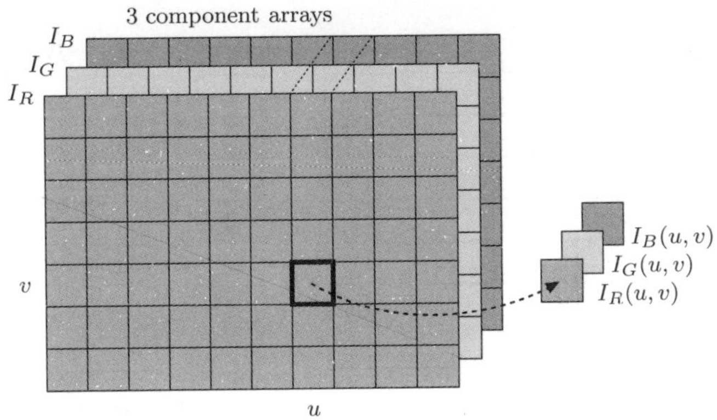

Figure 8.3 RGB color image in component ordering. The three color components are laid out in separate arrays I_R, I_G, I_B of the same size.

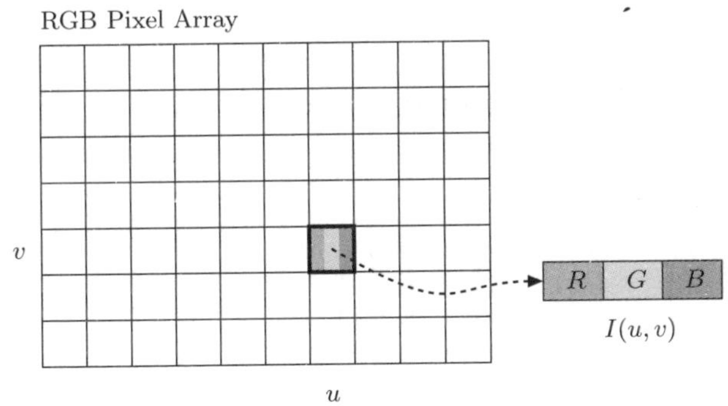

Figure 8.4 RGB-color image using packed ordering. The three color components R, G, and B are placed together in a single array element.

The access functions, Red(), Green(), Blue(), will depend on the specific implementation used for encoding the color pixels.

Indexed images

Indexed images permit only a limited number of distinct colors and therefore are used mostly for illustrations and graphics that contain large regions of the same color. Often these types of images are stored in indexed GIF or PNG files for use on the Web. In these indexed images, the pixel array does not contain color or brightness data but instead consists of integer numbers k that are used to index into a color table or "palette"

$$P(k) = (r_k, g_k, b_k),$$

for $k = 0 \ldots N{-}1$ (Fig. 8.5). N is the size of the color table and therefore also

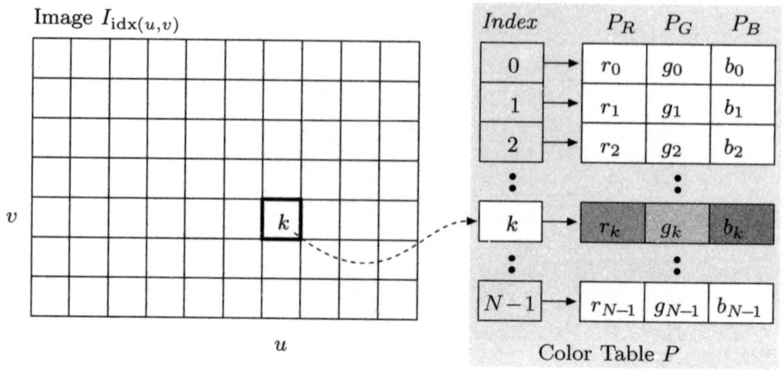

Figure 8.5 RGB indexed image. Instead of a full color value, each pixel in $I_{\text{idx}(u,v)}$ contains an index k. The color value for each k is defined by an entry in the color table or "palette" $P[k]$.

the maximum number of distinct image colors (typically $N = 2$ to 256). Since the color table can contain any RGB color value (r_k, g_k, b_k), it must be saved as part of the image. The RGB component values of an indexed image I_{idx} at location (u, v) are obtained as

$$\begin{pmatrix} R \\ G \\ B \end{pmatrix} \leftarrow \begin{pmatrix} P_R(k) \\ P_G(k) \\ P_B(k) \end{pmatrix} = \begin{pmatrix} r_k \\ g_k \\ b_k \end{pmatrix}, \quad \text{with } k = I_{\text{idx}}(u, v). \tag{8.3}$$

During the transformation from a true color image to an indexed image (for example, from a JPEG image to a GIF image), the problem of optimal color reduction, or color quantization, arises. Color quantization is the process of determining an optimal color table and then mapping it to the original colors. This process is described in detail in Vol. 2 [6, Sec. 5].

8.1.2 Color Images in ImageJ

ImageJ provides two simple types of color images:

- RGB full-color images (24-bit "RGB color")
- Indexed images ("8-bit color")

RGB true color images

RGB color images in ImageJ use a packed order (see Sec. 8.1.1), where each color pixel is represented by a 32-bit `int` value. As Fig. 8.6 illustrates, 8 bits are used to represent each of the RGB components, which limits the range of

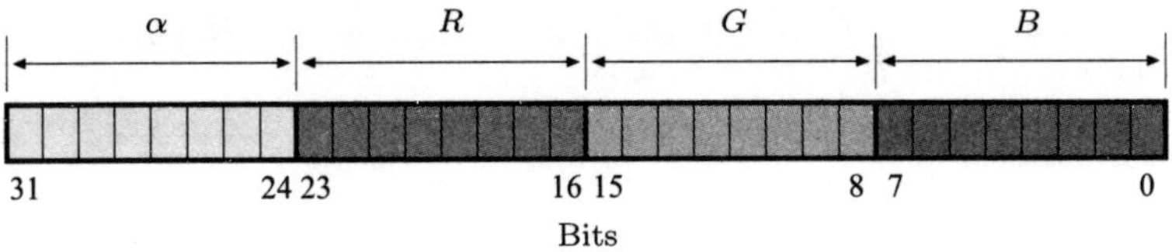

Figure 8.6 Structure of an RGB color pixel in Java. Within a 32-bit `int`, 8 bits are allocated, in the following order, for each of the color components R, G, B as well as the transparency α (unused in ImageJ).

the individual components to 0 to 255. The remaining 8 bits are reserved for the transparency,[2] or *alpha* (α), component. This is also the usual ordering in Java[3] for RGB color images.

Accessing RGB pixel values. RGB color images are represented by an array of pixels, the elements of which are standard Java `int`s. To disassemble the packed `int` value into the three color components, you apply the appropriate bitwise shifting and masking operations. In the following example, we assume that the image processor `ip` contains an RGB color image:

```
int c = ip.getPixel(u,v);          // a color pixel
int r = (c & 0xff0000) >> 16;      // red value
int g = (c & 0x00ff00) >> 8;       // green value
int b = (c & 0x0000ff);            // blue value
```

In this example, each of the RGB components of the packed pixel `c` are isolated using a bitwise AND operation (`&`) with an appropriate bit mask (following convention, bit masks are given in hexadecimal[4] notation), and afterwards the extracted bits are shifted right by 16 (for R) or 8 (for G) bit positions (see Fig. 8.7).

The "construction" of an RGB pixel from the individual R, G, and B values is done in the opposite direction using the bitwise OR operator (`|`) and shifting the bits left (`<<`):

```
int r = 169; // red value
int g = 212; // green value
int b = 17;  // blue value
int c = ((r & 0xff)<<16) | ((g & 0xff)<<8) | b & 0xff;
ip.putPixel(u,v,C);
```

Masking the component values with `0xff` works in this case because except for the bits in positions 0 to 7 (values in the range 0 to 255), all the other bits are

[2] The transparency value α (alpha) represents the ability to see through a color pixel onto the background. At this time, the α channel is unused in ImageJ.

[3] Java Advanced Window Toolkit (AWT).

[4] The mask `0xff0000` is of type `int` and represents the 32-bit binary pattern 00000000111111110000000000000000.

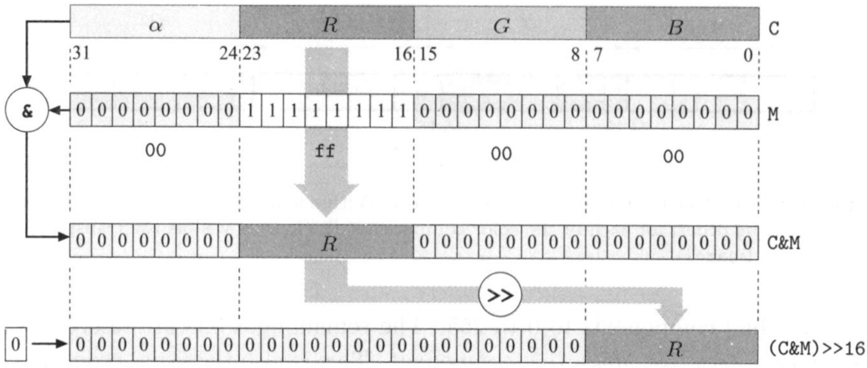

Figure 8.7 Decomposition of a 32-bit RGB color pixel using bit operations. The R component (bits 16–23) of the RGB pixels `C` (above) is isolated using a bitwise AND operation (`&`) together with a bit mask `M = 0xff0000`. All bits except the R component are set to the value 0, while the bit pattern within the R component remains unchanged. This bit pattern is subsequently shifted 16 positions to the right (`>>`), so that the R component is moved into the lowest 8 bits and its value lies in the range of 0 to 255. During the shift operation, zeros are filled in from the left.

already set to zero. A complete example of manipulating an RGB color image using bit operations is presented in Prog. 8.1. Instead of accessing color pixels using ImageJ's access functions, these programs directly access the pixel array for increased efficiency (see also Sec. B.1.3).

The ImageJ class `ColorProcessor` provides an easy to use alternative which returns the separated RGB components (as an `int` array with three elements). In the following example that demonstrates its use, `ip` is of type `ColorProcessor`:

```
int[] RGB = new int[3];
...
RGB = ip.getPixel(u,v,RGB);
int r = RGB[0];
int g = RGB[1];
int b = RGB[2];
...
ip.putPixel(u,v,RGB);
```

A more detailed and complete example is shown by the simple plugin in Prog. 8.2, which increases the value of all three color components of an RGB image by 10 units. Notice that the plugin limits the resulting component values to 255, because the `putPixel()` method only uses the lowest 8 bits of each component and does not test if the value passed in is out of the permitted 0 to 255 range. Without this test, arithmetic overflow errors can occur. The price for using this access method, instead of direct array access, is a noticeably longer running time (approximately a factor of 4 when compared to the version

```
 1 // File Brighten_Rgb_1.java
 2
 3 import ij.ImagePlus;
 4 import ij.plugin.filter.PlugInFilter;
 5 import ij.process.ImageProcessor;
 6
 7 public class Brighten_Rgb_1 implements PlugInFilter {
 8
 9   public int setup(String arg, ImagePlus im) {
10     return DOES_RGB; // this plugin works on RGB images
11   }
12
13   public void run(ImageProcessor ip) {
14     int[] pixels = (int[]) ip.getPixels();
15
16     for (int i = 0; i < pixels.length; i++) {
17       int c = pixels[i];
18       // split the color pixel into RGB components
19       int r = (c & 0xff0000) >> 16;
20       int g = (c & 0x00ff00) >> 8;
21       int b = (c & 0x0000ff);
22       // modify colors
23       r = r + 10; if (r > 255) r = 255;
24       g = g + 10; if (g > 255) g = 255;
25       b = b + 10; if (b > 255) b = 255;
26       // reassemble the color pixel and insert into pixel array
27       pixels[i]
28           = ((r & 0xff)<<16) | ((g & 0xff)<<8) | b & 0xff;
29     }
30   }
31
32 } // end of class Brighten_Rgb_1
```

Program 8.1 Working with RGB color images using bit operations (ImageJ plugin, version 1). This plugin increases the values of all three color components by 10 units. It demonstrates the use of direct access to the pixel array (line 17), the separation of color components using bit operations (lines 19–21), and the reassembly of color pixels after modification (line 28). The value `DOES_RGB` (defined in the interface `PlugInFilter`) returned by the `setup()` method indicates that this plugin is designed to work on RGB formatted true color images (line 10).

in Prog. 8.1).

Opening and saving RGB images. ImageJ supports the following types of image formats for RGB true color images:

- **TIFF** (only uncompressed): 3 × 8-bit RGB. TIFF color images with 16-bit depth are opened as an image stack consisting of three 16-bit intensity images.
- **BMP**, **JPEG**: 3 × 8-bit RGB.
- **PNG**: 3 × 8-bit RGB.

```java
1 // File Brighten_Rgb_2.java
2
3 import ij.ImagePlus;
4 import ij.plugin.filter.PlugInFilter;
5 import ij.process.ColorProcessor;
6 import ij.process.ImageProcessor;
7
8 public class Brighten_Rgb_2 implements PlugInFilter {
9   static final int R = 0, G = 1, B = 2; // component indices
10
11   public int setup(String arg, ImagePlus im) {
12     return DOES_RGB; // this plugin works on RGB images
13   }
14
15   public void run(ImageProcessor ip) {
16     //make sure the image is of type ColorProcessor
17     ColorProcessor cp = (ColorProcessor) ip;
18     int[] RGB = new int[3];
19
20     for (int v = 0; v < cp.getHeight(); v++) {
21       for (int u = 0; u < cp.getWidth(); u++) {
22         cp.getPixel(u, v, RGB);
23         RGB[R] = Math.min(RGB[R]+10, 255); // add 10 and
24         RGB[G] = Math.min(RGB[G]+10, 255); // limit to 255
25         RGB[B] = Math.min(RGB[B]+10, 255);
26         cp.putPixel(u, v, RGB);
27       }
28     }
29   }
30
31 } // end of class Brighten_Rgb_2
```

Program 8.2 Working with RGB color images without bit operations (ImageJ plugin, version 2). This plugin increases the values of all three color components by 10 units using the access methods `getPixel(int, int, int[])` and `putPixel(int, int, int[])` from the class `ColorProcessor` (lines 22 and 26, respectively). The running time, because of the method calls, is approximately four times higher than that of version 1 (Prog. 8.1).

- **RAW**: using the ImageJ menu File→Import→Raw, RGB images can be opened whose format is not directly supported by ImageJ. It is then possible to select different arrangements of the color components.

Creating RGB images. The simplest way to create a new RGB image using ImageJ is to use an instance of the class `ColorProcessor`, as the following example demonstrates:

```java
int w = 640, h = 480;
ColorProcessor cip = new ColorProcessor(w, h);
ImagePlus cimg = new ImagePlus("My New Color Image", cip);
cimg.show();
```

When needed, the color image can be displayed by creating an instance of the class `ImagePlus` (line 3) and calling its `show()` method. Since `cip` is of type `ColorProcessor`, the resulting `ImagePlus` object `cimg` is also a color image. The following code segment demonstrates how this could be verified:

```
5 if (cimg.getType()==ImagePlus.COLOR_RGB) {
6   int b = cimg.getBitDepth(); // b = 24
7   IJ.write("this is an RGB color image with " + b + " bits");
8 }
```

Indexed color images

The structure of an indexed image in ImageJ is given in Fig. 8.5, where each element of the index array is 8 bits and therefore can represent a maximum of 256 different colors. When programming, indexed images are similar to grayscale images, as both make use of a color table to determine the actual color of the pixel. Indexed images differ from grayscale images only in that the contents of the color table are not intensity values but RGB values.

Opening and saving indexed images. ImageJ supports the following types of image formats for indexed images:

- **GIF**: index values with 1 to 8 bits (2 to 256 colors), 3×8-bit color values.
- **PNG**: index values with 1 to 8 Bits (2 to 256 colors), 3×8-bit color values. When saved as PNG, indexed images are stored as full-color RGB images.
- **BMP**, **TIFF** (uncompressed): index values with 1 to 8 bits (2 to 256 colors), 3×8-bit color values.

Working with indexed images. The indexed format is mostly used as a space-saving means of image storage and is not directly useful as a processing format since an index value in the pixel array is arbitrarily related to the actual color, found in the color table, that it represents. When working with indexed images it usually makes no sense to base any numerical interpretations on the pixel values or to apply any filter operations designed for 8-bit intensity images. Figure 8.8 illustrates an example of applying a Gaussian filter and a median filter to the pixels of an indexed image. Since there is no meaningful quantitative relation between the actual colors and the index values, the results are erratic. Note that even the use of the median filter is inadmissible because no ordering relation exists between the index values. Thus, with few exceptions, ImageJ functions do not permit the application of such operations to indexed images. Generally, when processing an indexed image, you first convert it into a true color RGB image and then after processing convert it back into an indexed image.

```
// File Brighten_Index_Image.java

import ij.ImagePlus;
import ij.WindowManager;
import ij.plugin.filter.PlugInFilter;
import ij.process.ImageProcessor;
import java.awt.image.IndexColorModel;

public class Brighten_Index_Image implements PlugInFilter {

  public int setup(String arg, ImagePlus im) {
    return DOES_8C; // this plugin works on indexed color images
  }

  public void run(ImageProcessor ip) {
    IndexColorModel icm =
      (IndexColorModel) ip.getColorModel();
    int pixBits = icm.getPixelSize();
    int mapSize = icm.getMapSize();

    //retrieve the current lookup tables (maps) for R,G,B
    byte[] Rmap = new byte[mapSize]; icm.getReds(Rmap);
    byte[] Gmap = new byte[mapSize]; icm.getGreens(Gmap);
    byte[] Bmap = new byte[mapSize]; icm.getBlues(Bmap);

    //modify the lookup tables
    for (int idx = 0; idx < mapSize; idx++){
      int r = 0xff & Rmap[idx]; //mask to treat as unsigned byte
      int g = 0xff & Gmap[idx];
      int b = 0xff & Bmap[idx];
      Rmap[idx] = (byte) Math.min(r + 10, 255);
      Gmap[idx] = (byte) Math.min(g + 10, 255);
      Bmap[idx] = (byte) Math.min(b + 10, 255);
    }

    //create a new color model and apply to the image
    IndexColorModel icm2 =
      new IndexColorModel(pixBits, mapSize, Rmap, Gmap,Bmap);
    ip.setColorModel(icm2);

    //update the resulting image
    WindowManager.getCurrentImage().updateAndDraw();
  }

} // end of class Brighten_Index_Image
```

Program 8.3 Working with indexed images (ImageJ plugin). This plugin increases the brightness of an image by 10 units by modifying the image's color table (palette). The actual values in the pixel array, which are indices into the palette, are not changed.

(a) (b) (c)

Figure 8.8 Improper application of smoothing filters to an indexed color image. Indexed image with 16 colors (a) and results of applying a linear smoothing filter (b) and a 3×3 median filter (c) to the pixel array (that is, the *index* values). The application of a linear filter makes no sense, of course, since no meaningful relation exists between the index values in the pixel array and the actual image intensities. While the median filter (c) delivers seemingly plausible results in this case, its use is also inadmissible because no suitable ordering relation exists between the index values.

When an ImageJ plugin is supposed to process indexed images, its `setup()` method should return the `DOES_8C` ("8-bit color") flag. The plugin in Prog. 8.3 shows how to increase the intensity of the three color components of an indexed image by 10 units (analogously to Progs. 8.1 and 8.2 for RGB images). Notice how in indexed images only the palette is modified and the original pixel data, the index values, remain the same. The color table of `ImageProcessor` is accessible through a `ColorModel`[5] object, which can be read using the method `getColorModel()` and modified using `setColorModel()`.

The `ColorModel` object for indexed images (as well as 8-bit grayscale images) is a subtype of `IndexColorModel`, which contains three color tables (*maps*) representing the red, green, and blue components as separate `byte` arrays. The size of these tables (2 to 256) can be determined by calling the method `getMapSize()`. Note that the elements of the palette should be interpreted as *unsigned* bytes with values ranging from 0 to 255. Just as with

[5] Defined in the standard Java class `java.awt.image.ColorModel`.

grayscale pixel values, during the conversion to int values, these color component values must also be bitwise masked with 0xff as shown in Prog. 8.3 (lines 28–30).

As a further example, Prog. 8.4 shows how to convert an indexed image to a true color RGB image of type ColorProcessor. Conversion in this direction poses no problems because the RGB component values for a particular pixel are simply taken from the corresponding color table entry, as described by Eqn. (8.3). On the other hand, conversion in the other direction requires *quantization* of the RGB color space and is as a rule more difficult and involved (see Vol. 2 [6, Sec. 5] for more details). In practice, most applications make use of existing conversion methods such as those available in ImageJ (see pp. 200–200).

Creating indexed images. In ImageJ, no special method is provided for the creation of indexed images, so in almost all cases they are generated by converting an existing image. The following method demonstrates how to directly create an indexed image if required:

```
ByteProcessor makeIndexColorImage(int w, int h, int nColors) {
  // allocate red, green, blue color tables:
  byte[] Rmap = new byte[nColors];
  byte[] Gmap = new byte[nColors];
  byte[] Bmap = new byte[nColors];
  // color maps need to be filled here
  byte[] pixels = new byte[w * h];
  // pixel array (color indices) needs to be filled here
  IndexColorModel cm
    = new IndexColorModel(8, nColors, Rmap, Gmap, Bmap);
  return new ByteProcessor(w, h, pixels, cm);
}
```

The parameter nColors defines the number of colors (and thus the size of the palette) and must be a value in the range of 2 to 256. To use the above template, you would complete it with code that filled the three byte arrays for the RGB components (Rmap, Gmap, Bmap) and the index array (pixels) with the appropriate values.

Transparency. Transparency is one of the reasons indexed images are often used for Web graphics. In an indexed image, it is possible to define one of the index values so that it is displayed in a transparent manner and at selected image locations the background beneath the image shows through. In Java this can be controlled when creating the image's color model (IndexColorModel). As an example, to make color index 2 in Prog. 8.3 transparent, lines 37–39 would need to be modified as follows:

```
// File Index_To_Rgb.java

import ij.ImagePlus;
import ij.plugin.filter.PlugInFilter;
import ij.process.ColorProcessor;
import ij.process.ImageProcessor;
import java.awt.image.IndexColorModel;

public class Index_To_Rgb implements PlugInFilter {
  static final int R = 0, G = 1, B = 2;

  public int setup(String arg, ImagePlus im) {
    return DOES_8C + NO_CHANGES; //does not alter original image
  }

  public void run(ImageProcessor ip) {
    int w = ip.getWidth();
    int h = ip.getHeight();

    //retrieve the color table (palette) for R,G,B
    IndexColorModel icm =
        (IndexColorModel) ip.getColorModel();
    int mapSize = icm.getMapSize();
    byte[] Rmap = new byte[mapSize]; icm.getReds(Rmap);
    byte[] Gmap = new byte[mapSize]; icm.getGreens(Gmap);
    byte[] Bmap = new byte[mapSize]; icm.getBlues(Bmap);

    //create new 24-bit RGB image
    ColorProcessor cp = new ColorProcessor(w,h);
    int[] RGB = new int[3];
    for (int v = 0; v < h; v++) {
      for (int u = 0; u < w; u++) {
        int idx = ip.getPixel(u, v);
        RGB[R] = Rmap[idx];
        RGB[G] = Gmap[idx];
        RGB[B] = Bmap[idx];
        cp.set(u, v, RGB);
      }
    }
    ImagePlus cimg = new ImagePlus("RGB Image",cp);
    cimg.show();
  }

} // end of class Index_To_Rgb
```

Program 8.4 Converting an indexed image to a true color RGB image (ImageJ plugin).

```
int tIdx = 2; // index of transparent color
IndexColorModel icm2 = new
  IndexColorModel(pixBits, mapSize, Rmap, Gmap, Bmap, tIdx);
ip.setColorModel(icm2);
```

At this time, however, ImageJ does not support the transparency property; it is not considered during display, and it is lost when the image is saved.

Color image conversion in ImageJ

In ImageJ, the following methods for converting between different types of color and grayscale image objects of type `ImagePlus` and processor objects of type `ImageProcessor` are available:

Converting images of type `ImageProcessor`. ImageJ objects of type `ImageProcessor` can be converted using the methods listed in Table 8.1. Each of these methods returns a new `ImageProcessor` object, unless the original image is already of the desired type. If this is the case, only a reference to the original image processor is returned, i. e., *no* duplication or modification occurs. The following example demonstrates the conversion from an `ByteProcessor` (grayscale) image type to an RGB color image:

```
ByteProcessor ip1; // a grayscale image
...
ImageProcessor ip2 = ip1.convertToRGB();
// now ip2 is of type ColorProcessor, ip1 is unmodified.
...
```

In this case, a new object (`ip2`) of type `ColorProcessor` is created and the original object (`ip1`) remains unchanged.

Converting images of type `ImagePlus`. ImageJ image objects of type `ImagePlus` can be converted with the help of methods from the ImageJ class `ImageConverter`, as summarized in Table 8.2. The following example demonstrates the conversion to an RGB color image:

```
import ij.process.ImageConverter;
...
ImagePlus ipl;
...
ImageConverter ic = new ImageConverter(ipl);
ic.convertToRGB();
// ipl is an RGB image now
```

Note that the method `convertToRGB()` does not return a new image object, but instead modifies the original `ImagePlus` object `ipl`.

8.2 Color Spaces and Color Conversion

The RGB color system is well-suited for use in programming, as it is simple to manipulate and maps directly to the typical display hardware. When modifying

Table 8.1 Conversion methods for images of type `ImageProcessor`. If *doScaling* is `true` in the first two methods, the pixel values are automatically scaled to the maximum range of the new image.

Method	Description
`ImageProcessor convertToByte(boolean doScaling)`	Converts to an 8-bit grayscale image (`ByteProcessor`).
`ImageProcessor convertToShort(boolean doScaling)`	Converts to a 16-bit grayscale image (`ShortProcessor`).
`ImageProcessor convertToFloat()`	Converts to a 32-bit floating-point image (`FloatProcessor`).
`ImageProcessor convertToRGB()`	Converts to a 32-bit RGB color image (`ColorProcessor`).

Table 8.2 Methods of the ImageJ class `ImageConverter` for converting `ImagePlus` objects. Note that these methods do not create any new images, but instead modify the original `ImagePlus` object *ipl* used to instantiate the `ImageConverter`.

Method	Description
`ImageConverter(ImagePlus ipl)`	Instantiates an `ImageConverter` object for the image *ipl*.
`void convertToGray8()`	Converts *ipl* to an 8-bit grayscale image.
`void convertToGray16()`	Converts *ipl* to a 16-bit grayscale image.
`void convertToGray32()`	Converts *ipl* to a 32-bit grayscale image (`float`).
`void convertToRGB()`	Converts *ipl* to an RGB color image.
`void convertRGBtoIndexedColor(int nColors)`	Converts the RGB true color image *ipl* to an indexed image with 8-bit index values and *nColors* colors, performing color quantization.
`void convertToHSB()`	Converts *ipl* to a color image using the HSB color space (see Sec. 8.2.3).
`void convertHSBToRGB()`	Converts an HSB color space image *ipl* to an RGB color image.

colors within the RGB space, it is important to remember that the *metric*, or *measured distance* within this color space, does not proportionally correspond to our perception of color (e. g., doubling the value of the red component does not necessarily result in a color which appears to be twice as red). In general, in this space, modifying different color points by the same amount can cause very different changes in color. In addition, brightness changes in the RGB color space are also perceived as nonlinear.

Since any coordinate movement modifies color tone, saturation, and brightness all at once, color selection in RGB space is difficult and quite non-intuitive. Color selection is more intuitive in other color spaces, such as the HSV space (see Sec. 8.2.3), since perceptual color features, such as saturation, are represented individually and can be modified independently. Alternatives to the RGB color space are also used in applications such as the automatic separation of objects from a colored background (the *blue box* technique in television), encoding television signals for transmission, or in printing, and are thus also relevant in digital image processing.

Figure 8.9 shows the distribution of the colors from natural images in the RGB color space. The first half of this section introduces alternative color spaces and the methods of converting between them and later discusses the choices that need to be made to correctly convert a color image to grayscale. In addition to the classical color systems most widely used in programming, precise reference systems, such as the CIEXYZ color space, gain increasing importance in practical color processing.

8.2.1 Conversion to Grayscale

The conversion of an RGB color image to a grayscale image proceeds by computing the equivalent gray or *luminance* value Y for each RGB pixel. In its simplest form, Y could be computed as the average

$$Y = \mathrm{Avg}(R, G, B) = \frac{R+G+B}{3} \tag{8.4}$$

of the three color components R, G, and B. Since we perceive both red and green as being substantially brighter than blue, the resulting image will appear to be too dark in the red and green areas and too bright in the blue ones. Therefore, a weighted sum of the color components

$$Y = \mathrm{Lum}(R, G, B) = w_R \cdot R + w_G \cdot G + w_B \cdot B \tag{8.5}$$

is typically used to compute the equivalent luminance value. The weights most often used were originally developed for encoding analog color television signals (see Sec. 8.2.4):

$$w_R = 0.299\,, \qquad w_G = 0.587\,, \qquad w_B = 0.114\,. \tag{8.6}$$

(a)

(b)
RGB

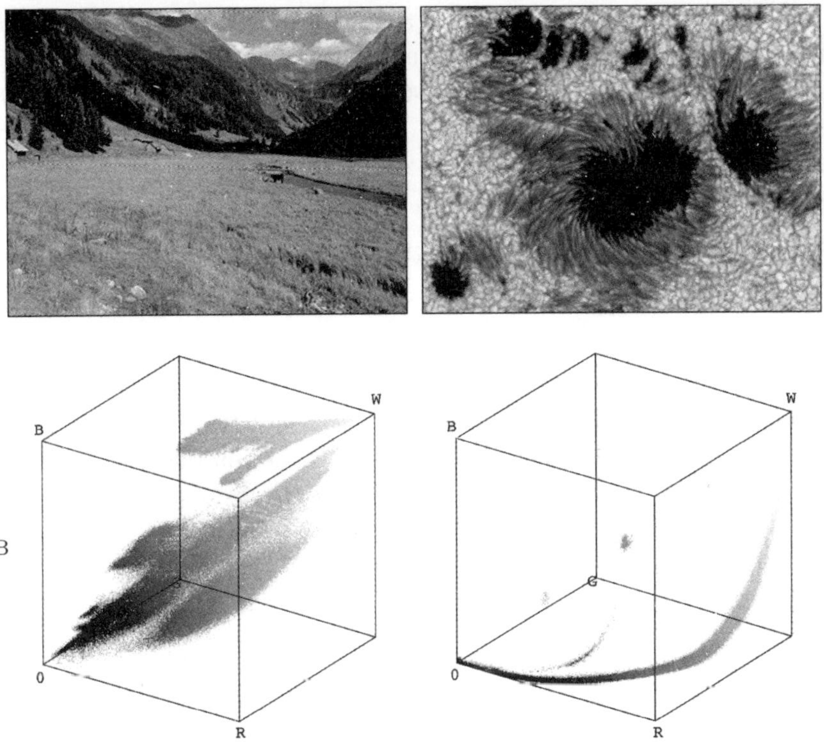

Figure 8.9 Examples of the color distribution of natural images. Original images: landscape photograph with dominant green and blue components and sun-spot image with rich red and yellow components (a). Distribution of image colors in RGB-space (b).

Those recommended in ITU-BT.709 [20] for digital color encoding are

$$w_R = 0.2125\,, \qquad w_G = 0.7154\,, \qquad w_B = 0.072\,. \tag{8.7}$$

If each color component is assigned the same weight, as in Eqn. (8.4), this is of course just a special case of Eqn. (8.5).

Note that, although these weights were developed for use with TV signals, they are optimized for *linear* RGB component values, i. e., signals with no gamma correction. In many practical situations, however, the RGB components are actually *nonlinear*, particularly when we work with sRGB images (see Vol. 2 [6, Sec. 6.3]). In this case, the RGB components must first be linearized to obtain the correct luminance values with the above weights. An alternative is to estimate the luminance without linearization by computing the weighted sum of the *nonlinear* component values and applying a different set of weights (for details see p. 109 of Vol. 2 [6, Sec. 6.3]).

In some color systems, instead of a weighted sum of the RGB color components, a nonlinear brightness function, for example the *value* V in HSV (Eqn. (8.11) in Sec. 8.2.3) or the *luminance* L in HLS (Eqn. (8.21)), is used as the intensity value Y.

Hueless (gray) color images

An RGB image is hueless or gray when the RGB components of each pixel $I(u, v) = (R, G, B)$ are the same; i. e., if

$$R = G = B.$$

Therefore, to completely remove the color from an RGB image, simply replace the R, G, and B component of each pixel with the equivalent gray value Y,

$$I_g(u, v) \leftarrow \begin{pmatrix} R_g \\ G_g \\ B_g \end{pmatrix} = \begin{pmatrix} Y \\ Y \\ Y \end{pmatrix}, \qquad (8.8)$$

by using $Y = \mathrm{Lum}(R, G, B)$ from Eqn. (8.5), for example. The resulting grayscale image should have the same subjective brightness as the original color image.

Grayscale conversion in ImageJ

In ImageJ, the simplest way to convert an RGB color image (of type `ColorProcessor`) into an 8-bit grayscale image is to use the method

```
convertToByte(boolean doScaling),
```

which returns a new image of type `ByteProcessor` (see Table 8.1 and the example on page 200). ImageJ uses the default weights $w_R = w_G = w_B = \frac{1}{3}$ (as in Eqn. (8.4)) for the RGB components, or $w_R = 0.299$, $w_G = 0.587$, $w_B = 0.114$ (as in Eqn. (8.6)) if the "Weighted RGB Conversions" option is selected in the Edit→Options→Conversions dialog. Arbitrary weights (*wr*, *wg*, *wb*) can be specified for subsequent conversion operations through the static `ColorProcessor` method

```
setWeightingFactors(double wr, double wg, double wb).
```

Similarly, the static method `ColorProcessor.getWeightingFactors()` can be used to retrieve the current weights as a 3-element `double`-array. Note that no linearization is performed on the color components, which should be considered when working with (nonlinear) sRGB colors (see Vol. 2 [6, Sec. 6.3] for details).

8.2.2 Desaturating Color Images

Desaturation is the uniform reduction of the amount of color in an RGB image in a *continuous* manner. It is done by replacing each RGB pixel by a desaturated color (R_d, G_d, B_d) computed by linear interpolation between the pixel's original color and the corresponding (Y, Y, Y) gray point in the RGB space, i. e.,

$$\begin{pmatrix} R_d \\ G_d \\ B_d \end{pmatrix} \leftarrow \begin{pmatrix} Y \\ Y \\ Y \end{pmatrix} + s_{\mathrm{col}} \cdot \begin{pmatrix} R-Y \\ G-Y \\ B-Y \end{pmatrix}, \tag{8.9}$$

where the factor $s_{\mathrm{col}} \in [0, 1]$ controls the remaining amount of color saturation (Fig. 8.10). A value of $s_{\mathrm{col}} = 0$ completely eliminates all color, resulting in a true grayscale image, and with $s_{\mathrm{col}} = 1$ the color values will be unchanged. In Prog. 8.5, continuous desaturation as defined in Eqn. (8.9) is implemented as an ImageJ plugin.

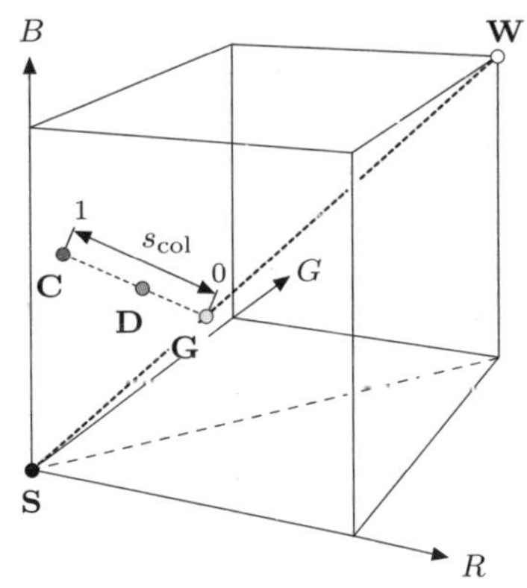

Figure 8.10 Desaturation in RGB space: original color point $\mathbf{C} = (R, G, B)$, its corresponding gray point $\mathbf{G} = (Y, Y, Y)$, and the desaturated color point $\mathbf{D} = (R_d, G_d, B_d)$. Saturation is controlled by the factor s_{col}.

8.2.3 HSV/HSB and HLS Color Space

In the **HSV** color space, colors are specified by the components *hue*, *saturation*, and *value*. Often, such as in Adobe products and the Java API, the **HSV** space is called **HSB**. While the acronym is different (in this case B = *brightness*),[6] it denotes the same color space. The HSV color space is traditionally shown

[6] Sometimes the HSV space is also referred to as the "HSI" space, where 'I' stands for *intensity*.

```
 1 // File Desaturate_Rgb.java
 2
 3 import ij.ImagePlus;
 4 import ij.plugin.filter.PlugInFilter;
 5 import ij.process.ImageProcessor;
 6
 7 public class Desaturate_Rgb implements PlugInFilter {
 8
 9   static double sCol = 0.3; // color saturation factor
10
11   public int setup(String arg, ImagePlus im) {
12     return DOES_RGB;
13   }
14
15   public void run(ImageProcessor ip) {
16
17     // iterate over all pixels
18     for (int v = 0; v < ip.getHeight(); v++) {
19       for (int u = 0; u < ip.getWidth(); u++) {
20
21         // get int-packed color pixel
22         int c = ip.get(u, v);
23
24         // extract RGB components from color pixel
25         int r = (c & 0xff0000) >> 16;
26         int g = (c & 0x00ff00) >> 8;
27         int b = (c & 0x0000ff);
28
29         // compute equivalent gray value
30         double y = 0.299 * r + 0.587 * g + 0.114 * b;
31
32         // linearly interpolate (yyy) ↔ (rgb)
33         r = (int) (y + sCol * (r - y));
34         g = (int) (y + sCol * (g - y));
35         b = (int) (y + sCol * (b - y));
36
37         // reassemble color pixel
38         c = ((r & 0xff)<<16) | ((g & 0xff)<<8) | b & 0xff;
39         ip.set(u, v, c);
40       }
41     }
42   }
43
44 } // end of class Desaturate_Rgb
```

Program 8.5 Continuous desaturation of an RGB color image (ImageJ plugin). The amount of color saturation is controlled by the variable `sCol` defined in line 9 (see Eqn. (8.9)).

as an upside-down, six-sided pyramid (Fig. 8.11 (a)), where the vertical axis represents the V (brightness) value, the horizontal distance from the axis the S (saturation) value, and the angle the H (hue) value. The black point is at

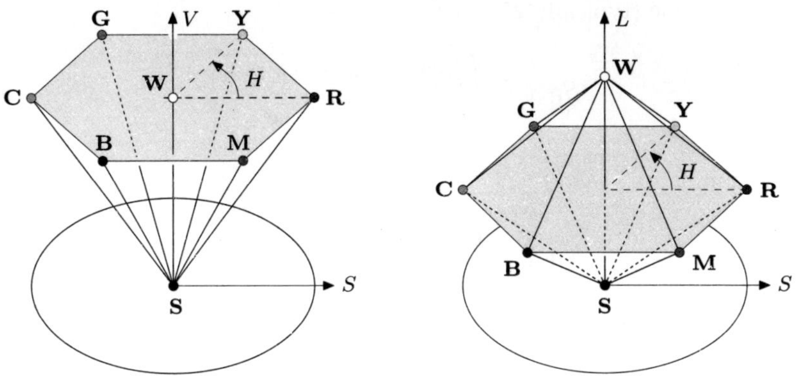

(a) HSV (b) HLS

Figure 8.11 HSV and HLS color space are traditionally visualized as a single or double hexagonal pyramid. The brightness V (or L) is represented by the vertical dimension, the color saturation S by the radius from the pyramid's axis, and the hue h by the angle. In both cases, the primary colors red (**R**), green (**G**), and blue (**B**) and the mixed colors yellow (**Y**), cyan (**C**), and magenta (**M**) lie on a common plane with black (**S**) at the tip. The essential difference between the HSV and HLS color spaces is the location of the white point (**W**).

the tip of the pyramid and the white point lies in the center of the base. The three primary colors *red*, *green*, and *blue* and the pairwise mixed colors *yellow*, *cyan* and *magenta* are the corner points of the base. While this space is often represented as a pyramid, according to its mathematical definition, the space is actually a *cylinder*, as shown below (Fig. 8.13).

The **HLS** color space[7] (*hue*, *luminance*, *saturation*) is very similar to the HSV space, and the *hue* component is in fact completely identical in both spaces. The *luminance* and *saturation* values also correspond to the vertical axis and the radius, respectively, but are defined differently than in HSV space. The common representation of the HLS space is as a double pyramid (Fig. 8.11 (b)), with black on the bottom tip and white on the top. The primary colors lie on the corner points of the hexagonal base between the two pyramids. Even though it is often portrayed in this intuitive way, mathematically the HLS space is again a cylinder (see Fig. 8.15).

RGB→HSV

To convert from RGB to the HSV color space, we first find the *saturation* of the RGB color components $R, G, B \in [0, C_{\max}]$, with $C_{\max}$ being the maximum

[7] The acronyms HLS and HSL are used interchangeably.

component value (typically 255), as

$$S_{\mathrm{HSV}} = \begin{cases} \frac{C_{\mathrm{rng}}}{C_{\mathrm{high}}} & \text{for } C_{\mathrm{high}} > 0 \\ 0 & \text{otherwise,} \end{cases} \tag{8.10}$$

and the luminance (*value*)

$$V_{\mathrm{HSV}} = \frac{C_{\mathrm{high}}}{C_{\max}}, \tag{8.11}$$

with C_{high}, C_{low}, and C_{rng} defined as

$$C_{\mathrm{high}} = \max(R, G, B), \quad C_{\mathrm{low}} = \min(R, G, B), \quad C_{\mathrm{rng}} = C_{\mathrm{high}} - C_{\mathrm{low}}. \tag{8.12}$$

Finally, we need to specify the *hue* value H_{HSV}. When all three RGB color components have the same value ($R = G = B$), then we are dealing with an *achromatic* (gray) pixel. In this particular case $C_{\mathrm{rng}} = 0$ and thus the saturation value $S_{\mathrm{HSV}} = 0$, consequently the hue is undefined. To compute H_{HSV} when $C_{\mathrm{rng}} > 0$, we first normalize each component using

$$R' = \frac{C_{\mathrm{high}} - R}{C_{\mathrm{rng}}}, \qquad G' = \frac{C_{\mathrm{high}} - G}{C_{\mathrm{rng}}}, \qquad B' = \frac{C_{\mathrm{high}} - B}{C_{\mathrm{rng}}}. \tag{8.13}$$

Then, depending on which of the three original color components had the maximal value, we compute a preliminary hue H' as

$$H' = \begin{cases} B' - G' & \text{if } R = C_{\mathrm{high}} \\ R' - B' + 2 & \text{if } G = C_{\mathrm{high}} \\ G' - R' + 4 & \text{if } B = C_{\mathrm{high}}. \end{cases} \tag{8.14}$$

Since the resulting value for H' lies on the interval $[-1 \ldots 5]$, we obtain the final hue value by normalizing to the interval $[0, 1]$ as

$$H_{\mathrm{HSV}} = \frac{1}{6} \cdot \begin{cases} (H' + 6) & \text{for } H' < 0 \\ H' & \text{otherwise.} \end{cases} \tag{8.15}$$

Hence all three components H_{HSV}, S_{HSV}, and V_{HSV} will lie within the interval $[0, 1]$. The hue value H_{HSV} can naturally also be computed in another angle interval, for example in the 0 to 360° interval using

$$H^{\circ}_{\mathrm{HSV}} = H_{\mathrm{HSV}} \cdot 360.$$

Under this definition, the RGB space unit cube is mapped to a *cylinder* with height and radius of length 1 (Fig. 8.13). In contrast to the traditional representation (Fig. 8.11), all HSB points within the entire cylinder correspond to valid color coordinates in RGB space. The mapping from RGB to the HSV space is nonlinear, as can be noted by examining how the black point stretches

H_{HSV} S_{HSV} V_{HSV}

Figure 8.12 HSV components for the test image in Fig. 8.2. The darker areas in the h_{HSV} component correspond to the red and yellow colors, where the *hue* angle is near zero.

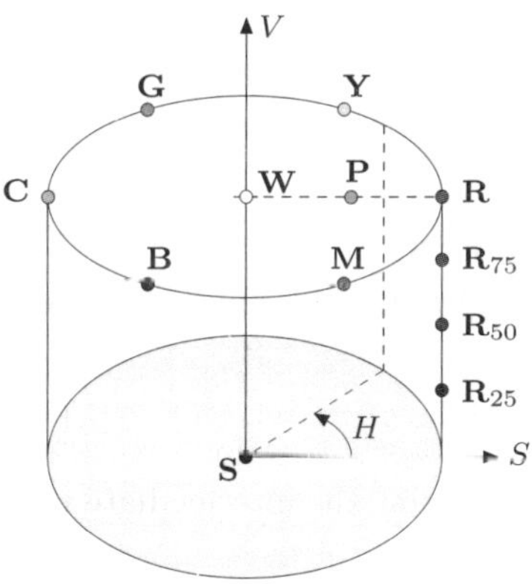

RGB/HSV Values

Pt.	Color	R	G	B	H	S	V
S	Black	0.00	0.00	0.00	—	0.00	0.00
R	Red	1.00	0.00	0.00	0	1.00	1.00
Y	Yellow	1.00	1.00	0.00	1/6	1.00	1.00
G	Green	0.00	1.00	0.00	2/6	1.00	1.00
C	Cyan	0.00	1.00	1.00	3/6	1.00	1.00
B	Blue	0.00	0.00	1.00	4/6	1.00	1.00
M	Magenta	1.00	0.00	1.00	5/6	1.00	1.00
W	White	1.00	1.00	1.00	—	0.00	1.00
$\mathbf{R}_{75}$	75% Red	0.75	0.00	0.00	0	1.00	0.75
$\mathbf{R}_{50}$	50% Red	0.50	0.00	0.00	0	1.00	0.50
$\mathbf{R}_{25}$	25% Red	0.25	0.00	0.00	0	1.00	0.25
P	Pink	1.00	0.50	0.50	0	0.5	1.00

Figure 8.13 HSV color space. The illustration shows the HSV color space as a cylinder with the coordinates H (*hue*) as the angle, S (*saturation*) as the radius, and V (*brightness value*) as the distance along the vertical axis, which runs between the black point **S** and the white point **W**. The table lists the (R, G, B) and (H, S, V) values of the color points marked on the graphic. Pure colors (composed of only one or two components) lie on the outer wall of the cylinder ($S = 1$), as exemplified by the gradually saturated reds ($\mathbf{R}_{25}$, $\mathbf{R}_{50}$, $\mathbf{R}_{75}$, $\mathbf{R}$).

completely across the cylinder's base. Figure 8.12 shows the individual HSV components (in grayscale) of the test image in Fig. 8.2. Figure 8.13 plots the location of some notable color points and compares them with their locations in RGB space (see also Fig. 8.1).

Java implementation. In Java, the RGB-HSV conversion is implemented in the class `java.awt.Color` by the method

```
float[] RGBtoHSB (int r, int g, int b, float[] hsv)
```

(HSV and HSB denote the same color space). The method takes three `int` arguments `r`, `g`, `b` (within the range $[0, 255]$) and returns a `float` array with the resulting H, S, V values in the interval $[0, 1]$. When an existing `float` array is passed as the argument *hsv*, then the result is placed in it; otherwise (when *hsv* = `null`) a new array is created. Here is a simple example:

```
import java.awt.Color;
...
float[] hsv = new float[3];
int red = 128, green = 255, blue = 0;
hsv = Color.RGBtoHSB (red, green, blue, hsv);
float h = hsv[0];
float s = hsv[1];
float v = hsv[2];
...
```

A possible implementation of the Java method `RGBtoHSB()` using the definition in Eqns. (8.11)–(8.15) is given in Prog. 8.6.

HSV→RGB

To convert an HSV tuple $(H_{\mathrm{HSV}}, S_{\mathrm{HSV}}, V_{\mathrm{HSV}})$, where H_{HSV}, S_{HSV}, and V_{HSV} $\in [0, 1]$, into the corresponding (R, G, B) color values, the appropriate color sector

$$H' = (6 \cdot H_{\mathrm{HSV}}) \bmod 6 \tag{8.16}$$

$(0 \leq H' < 6)$ is determined first, followed by computing the intermediate values

$$\begin{aligned} c_1 &= \lfloor H' \rfloor, & x &= (1 - S_{\mathrm{HSV}}) \cdot v, \\ c_2 &= H' - c_1, & y &= (1 - (S_{\mathrm{HSV}} \cdot c_2)) \cdot V_{\mathrm{HSV}}, \\ & & z &= (1 - (S_{\mathrm{HSV}} \cdot (1 - c_2))) \cdot V_{\mathrm{HSV}}. \end{aligned} \tag{8.17}$$

Depending on the value of c_1, the normalized RGB values $R', G', B' \in [0, 1]$ are then computed from $v = V_{\mathrm{HSV}}$, x, y, and z as follows:[8]

$$(R', G', B') = \begin{cases} (v, z, x) & \text{if } c_1 = 0 \\ (y, v, x) & \text{if } c_1 = 1 \\ (x, v, z) & \text{if } c_1 = 2 \\ (x, y, v) & \text{if } c_1 = 3 \\ (z, x, v) & \text{if } c_1 = 4 \\ (v, x, y) & \text{if } c_1 = 5. \end{cases} \tag{8.18}$$

[8] The variables x, y, z used here have no relation to those used in the CIEXYZ color space.

```
static float[] RGBtoHSV (int R, int G, int B, float[] HSV) {
  // R, G, B ∈ [0, 255]
  float H = 0, S = 0, V = 0;
  float cMax = 255.0f;
  int cHi = Math.max(R,Math.max(G,B)); // highest color value
  int cLo = Math.min(R,Math.min(G,B)); // lowest color value
  int cRng = cHi - cLo;          // color range

  // compute value V
  V = cHi / cMax;

  // compute saturation S
  if (cHi > 0)
    S = (float) cRng / cHi;

  // compute hue H
  if (cRng > 0) { // hue is defined only for color pixels
    float rr = (float)(cHi - R) / cRng;
    float gg = (float)(cHi - G) / cRng;
    float bb = (float)(cHi - B) / cRng;
    float hh;
    if (R == cHi)                      // R is highest color value
      hh = bb - gg;
    else if (G == cHi)                 // G is highest color value
      hh = rr - bb + 2.0f;
    else                               // B is highest color value
      hh = gg - rr + 4.0f;
    if (hh < 0)
      hh= hh + 6;
    H = hh / 6;
  }

  if (HSV == null) // create a new HSV array if needed
    HSV = new float[3];
  HSV[0] = H; HSV[1] = S; HSV[2] = V;
  return HSV;
}
```

Program 8.6 RGB→HSV conversion. This Java method for RGB→HSV conversion follows the process given in the text to compute a single color tuple. It takes the same arguments and returns results identical to the standard `Color.RGBtoHSB()` method.

The scaling of the RGB components to whole numbers in the range $[0, N-1]$ (typically $N = 256$) is carried out as follows:

$$\begin{aligned} R &= \min\bigl(\operatorname{round}(N \cdot R'), N-1\bigr), \\ G &= \min\bigl(\operatorname{round}(N \cdot G'), N-1\bigr), \\ B &= \min\bigl(\operatorname{round}(N \cdot B'), N-1\bigr). \end{aligned} \tag{8.19}$$

```
static int HSVtoRGB (float h, float s, float v) {
  // h, s, v ∈ [0,1]
  float rr = 0, gg = 0, bb = 0;
  float hh = (6 * h) % 6;                    // h' ← (6 · h) mod 6
  int   c1 = (int) hh;                       // c1 ← ⌊h'⌋
  float c2 = hh - c1;
  float x = (1 - s) * v;
  float y = (1 - (s * c2)) * v;
  float z = (1 - (s * (1 - c2))) * v;
  switch (c1) {
    case 0: rr=v; gg=z; bb=x; break;
    case 1: rr=y; gg=v; bb=x; break;
    case 2: rr=x; gg=v; bb=z; break;
    case 3: rr=x; gg=y; bb=v; break;
    case 4: rr=z; gg=x; bb=v; break;
    case 5: rr=v; gg=x; bb=y; break;
  }
  int N = 256;
  int r = Math.min(Math.round(rr*N), N-1);
  int g = Math.min(Math.round(gg*N), N-1);
  int b = Math.min(Math.round(bb*N), N-1);
  // create int-packed RGB color:
  int rgb = ((r&0xff)<<16) | ((g&0xff)<<8) | b&0xff;
  return rgb;
}
```

Program 8.7 HSV→RGB conversion. This Java method takes the same arguments and returns identical results as the standard method `Color.HSBtoRGB()`.

Java implementation. In Java, HSV→RGB conversion is implemented in the standard AWT class `java.awt.Color` by the method

`int HSBtoRGB (float h, float s, float v)` ,

which takes three `float` arguments $h, s, v \in [0, 1]$ and returns the corresponding RGB color as an `int` value with 3×8 bits arranged in the standard Java RGB format (see Fig. 8.6). One possible implementation of this method is shown in Prog. 8.7.

RGB→HLS

In the HLS model, the *hue* value H_{HLS} is computed in the same way as in the HSV model (Eqns. (8.13)–(8.15)), i. e.,

$$H_{\mathrm{HLS}} = H_{\mathrm{HSV}}. \tag{8.20}$$

The other values, L_{HLS} and S_{HLS}, are computed as follows (for C_{high}, C_{low}, and C_{rng}, see Eqn. (8.12)):

$$L_{\mathrm{HLS}} = \frac{C_{\mathrm{high}} + C_{\mathrm{low}}}{2}, \tag{8.21}$$

Figure 8.14 HLS color components H_{HLS} (*hue*), S_{HLS} (*saturation*), and L_{HLS} (*luminance*). Note that the S and L images are swapped to appear in the same order as in HSV space (see Fig. 8.12).

$$S_{\text{HLS}} = \begin{cases} 0 & \text{for } L_{\text{HLS}} = 0 \\ 0.5 \cdot \frac{C_{\text{rng}}}{L_{\text{HLS}}} & \text{for } 0 < L_{\text{HLS}} \leq 0.5 \\ 0.5 \cdot \frac{C_{\text{rng}}}{1-L_{\text{HLS}}} & \text{for } 0.5 < L_{\text{HLS}} < 1 \\ 0 & \text{for } L_{\text{HLS}} = 1. \end{cases} \tag{8.22}$$

Figure 8.14 shows the individual HLS components of the test image as grayscale images. Using the above definitions, the unit cube in the RGB space is again mapped to a cylinder with height and length 1 (Fig. 8.15). In contrast to the HSV space (Fig. 8.13), the primary colors lie together in the horizontal plane at $L_{\text{HLS}} = 0.5$ and the white point lies outside of this plane at $L_{\text{HLS}} = 1.0$. Using these nonlinear transformations, the black and the white points are mapped to the top and the bottom planes of the cylinder, respectively.

HLS→RGB

When converting from HLS to the RGB space, we assume that H_{HLS}, S_{HLS}, $L_{\text{HLS}} \in [0, 1]$. In the case where $L_{\text{HLS}} = 0$ or $L_{\text{HLS}} = 1$, the result is

$$(R', G', B') = \begin{cases} (0, 0, 0) & \text{for } L_{\text{HLS}} = 0 \\ (1, 1, 1) & \text{for } L_{\text{HLS}} = 1. \end{cases} \tag{8.23}$$

Otherwise, we again determine the appropriate color sector

$$H' = (6 \cdot H_{\text{HLS}}) \bmod 6\,, \tag{8.24}$$

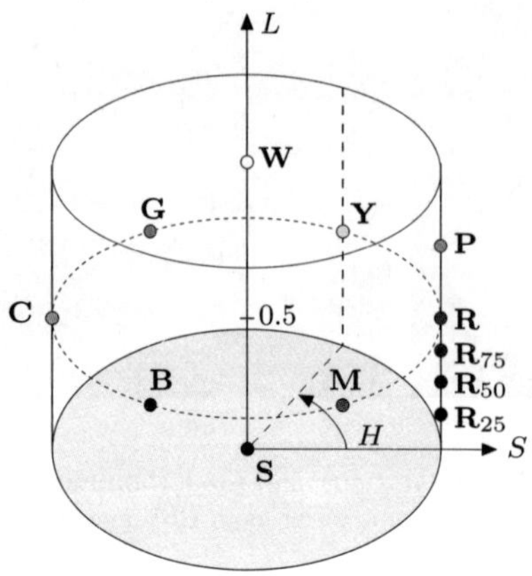

RGB/HLS Values

Pt.	Color	R	G	B	H	S	L
S	Black	0.00	0.00	0.00	—	0.00	0.00
R	Red	1.00	0.00	0.00	0	1.00	0.50
Y	Yellow	1.00	1.00	0.00	1/6	1.00	0.50
G	Green	0.00	1.00	0.00	2/6	1.00	0.50
C	Cyan	0.00	1.00	1.00	3/6	1.00	0.50
B	Blue	0.00	0.00	1.00	4/6	1.00	0.50
M	Magenta	1.00	0.00	1.00	5/6	1.00	0.50
W	White	1.00	1.00	1.00	—	0.00	1.00
$\mathbf{R}_{75}$	75% Red	0.75	0.00	0.00	0	1.00	0.375
$\mathbf{R}_{50}$	50% Red	0.50	0.00	0.00	0	1.00	0.250
$\mathbf{R}_{25}$	25% Red	0.25	0.00	0.00	0	1.00	0.125
P	Pink	1.00	0.50	0.50	0/6	1.00	0.75

Figure 8.15 HLS color space. The illustration shows the HLS color space visualized as a cylinder with the coordinates H (*hue*) as the angle, S (*saturation*) as the radius, and L (*lightness*) as the distance along the vertical axis, which runs between the black point **S** and the white point **W**. The table lists the (R, G, B) and (H, S, L) values where "pure" colors (created using only one or two color components) lie on the lower half of the outer cylinder wall ($S = 1$), as illustrated by the gradually saturated reds ($\mathbf{R}_{25}$, $\mathbf{R}_{50}$, $\mathbf{R}_{75}$, $\mathbf{R}$). Mixtures of all three primary colors, where at least one of the components is completely saturated, lie along the upper half of the outer cylinder wall; for example, the point **P** (pink).

where $(0 \leq H' < 6)$, and then, based on the resulting sector, we determine the values

$$
\begin{aligned}
c_1 &= \lfloor H' \rfloor & d &= \begin{cases} S_{\mathrm{HLS}} \cdot L_{\mathrm{HLS}} & \text{for } L_{\mathrm{HLS}} \leq 0.5 \\ S_{\mathrm{HLS}} \cdot (L_{\mathrm{HLS}} - 1) & \text{for } L_{\mathrm{HLS}} > 0.5 \end{cases} \\
c_2 &= H' - c_1 & & \\
w &= L_{\mathrm{HLS}} + d & y &= w - (w - x) \cdot c_2 \\
x &= L_{\mathrm{HLS}} - d & z &= x + (w - x) \cdot c_2 .
\end{aligned}
\tag{8.25}
$$

The assignment of the RGB values is done similarly to Eqn. (8.18), i. e.,

$$
(R', G', B') = \begin{cases} (w, z, x) & \text{if } c_1 = 0 \\ (y, w, x) & \text{if } c_1 = 1 \\ (x, w, z) & \text{if } c_1 = 2 \\ (x, y, w) & \text{if } c_1 = 3 \\ (z, x, w) & \text{if } c_1 = 4 \\ (w, x, y) & \text{if } c_1 = 5. \end{cases}
\tag{8.26}
$$

Finally, scaling the normalized ($[0, 1]$) R', G', B' color components back into the $[0, 255]$ range is done as in Eqn. (8.19).

```java
static float[] RGBtoHLS (float R, float G, float B) {
  // R, G, B assumed to be in [0, 1]
  float cHi = Math.max(R,Math.max(G,B)); // highest color value
  float cLo = Math.min(R,Math.min(G,B)); // lowest color value
  float cRng = cHi - cLo;         // color range

  // compute luminance L
  float L = (cHi + cLo)/2;

  // compute saturation S
  float S = 0;
  if (0 < L && L < 1) {
    float d = (L <= 0.5f) ? L : (1 - L);
    S = 0.5f * cRng / d;
  }

  // compute hue H
  float H=0;
  if (cHi > 0 && cRng > 0) {      // a color pixel
    float rr = (float)(cHi - R) / cRng;
    float gg = (float)(cHi - G) / cRng;
    float bb = (float)(cHi - B) / cRng;
    float hh;
    if (R == cHi)                     // R is highest color value
      hh = bb - gg;
    else if (G == cHi)                // G is highest color value
      hh = rr - bb + 2.0f;
    else                              // B is highest color value
      hh = gg - rr + 4.0f;

    if (hh < 0)
      hh= hh + 6;
    H = hh / 6;
  }

  return new float[] {H,L,S};
}
```

Program 8.8 RGB→HLS conversion (Java method).

Java implementation (RGB ↔ HLS)

Currently there is no method in either the standard Java API or ImageJ for converting color values between RGB and HLS. Program 8.8 gives one possible implementation of the RGB→HLS conversion that follows the definitions in Eqns. (8.20)–(8.22). The HLS→RGB conversion is given in Prog. 8.9.

```
static float[] HLStoRGB (float H, float L, float S) {
  // H, L, S assumed to be in [0, 1]
  float R = 0, G = 0, B =0;

  if (L <= 0)       // black
    R = G = B = 0;
  else if (L >= 1)  // white
    R = G = B = 1;
  else {
    float hh = (6 * H) % 6;
    int   c1 = (int) hh;
    float c2 = hh - c1;
    float d = (L <= 0.5f) ? (S * L) : (S * (1 - L));
    float w = L + d;
    float x = L - d;
    float y = w - (w - x) * c2;
    float z = x + (w - x) * c2;
    switch (c1) {
      case 0: R=w; G=z; B=x; break;
      case 1: R=y; G=w; B=x; break;
      case 2: R=x; G=w; B=z; break;
      case 3: R=x; G=y; B=w; break;
      case 4: R=z; G=x; B=w; break;
      case 5: R=w; G=x; B=y; break;
    }
  }
  return new float[] {R,G,B};
}
```

Program 8.9 HLS→RGB conversion (Java method).

Comparing HSV and HLS

Despite the gross similarity between the two color spaces, as Fig. 8.16 illustrates, substantial differences in the V/L and S components do exist. The essential difference between the HSV and HLS spaces is the ordering of the colors that lie between the white point **W** and the "pure" colors (**R**, **G**, **B**, **Y**, **C**, **M**), which consist of at most two primary colors, at least one of which is completely saturated.

The difference in how colors are distributed in RGB, HSV, and HLS space is readily apparent in Fig. 8.17. The starting point was a distribution of 1331 ($11 \times 11 \times 11$) color tuples obtained by uniformly sampling the RGB space at an interval of 0.1 in each dimension. The color distributions in HSV-space for a set of natural images are shown in Fig. 8.18 (p. 220).

Both the HSV and HLS color spaces are widely used in practice; for instance, for selecting colors in image editing and graphics design applications. In digital image processing, they are also used for *color keying* (that is, isolating objects according to their *hue*) on a homogeneously colored background where the

Figure 8.16 Comparison between HSV and HLS components: *saturation* (top row) and *intensity* (bottom row). In the color *saturation* difference image $S_{\mathrm{HSV}} - S_{\mathrm{HLS}}$ (top), light areas correspond to positive values and dark areas to negative values. Saturation in the HLS representation, especially in the brightest sections of the image, is notably higher, resulting in negative values in the difference image. For the *intensity* (*value* and *luminance*, respectively) in general, $V_{\mathrm{HSV}} \geq L_{\mathrm{HLS}}$ and therefore the difference $V_{\mathrm{HSV}} - L_{\mathrm{HLS}}$ (bottom) is always positive. The *hue* component H (not shown) is identical in both representations.

brightness is not necessarily constant.

8.2.4 TV Color Spaces—YUV, YIQ, and $\mathrm{YC_bC_r}$

These color spaces are an integral part of the standards surrounding the recording, storage, transmission, and display of television signals. YUV and YIQ are the fundamental color-encoding methods for the analog NTSC and PAL systems, and $\mathrm{YC_bC_r}$ is a part of the international standards governing digital television [19]. All of these color spaces have in common the idea of separating the luminance component Y from two chroma components and, instead of directly encoding colors, encoding color differences. In this way, compatibility with legacy black and white systems is maintained while at the same time the bandwidth of the signal can be optimized by using different transmission bandwidths for the brightness and the color components. Since the human

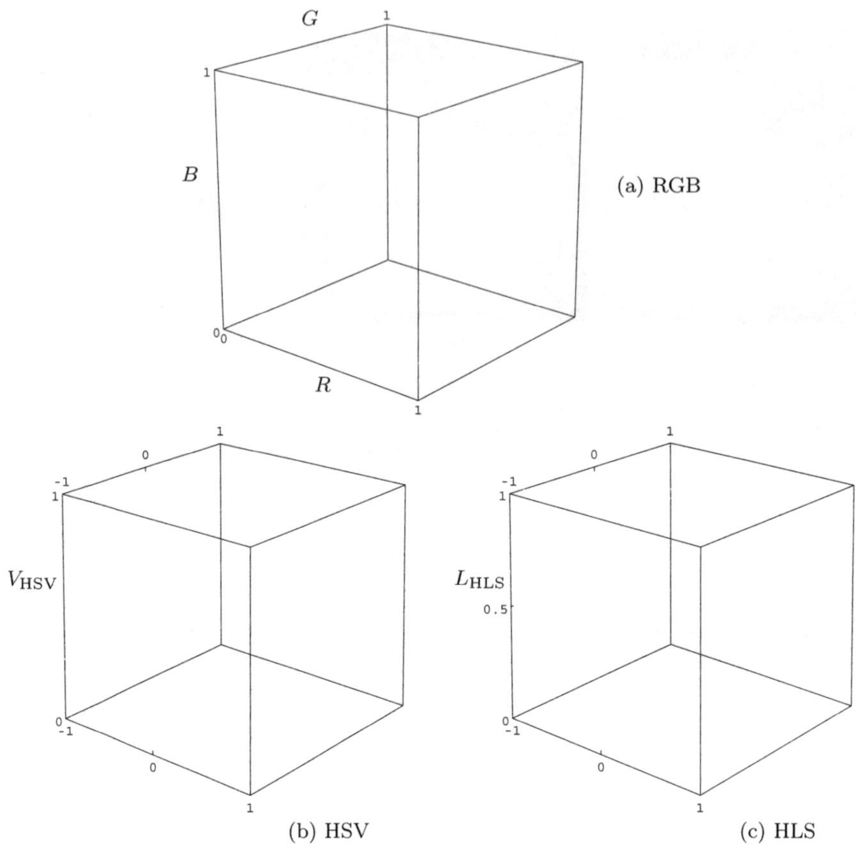

Figure 8.17 Distribution of colors in the RGB, HSV, and HLS spaces. The starting point is the uniform distribution of colors in RGB space (a). The corresponding colors in the cylindrical spaces are distributed nonsymmetrically in HSV (b) and symmetrically in HLS (c).

visual system is not able to perceive detail in the color components as well as it does in the intensity part of a video signal, the amount of information, and consequently bandwidth, used in the color channel can be reduced to approximately 1/4 of that used for the intensity component. This fact is also used when compressing digital still images and is why, for example, the JPEG codec converts RGB images to YC_bC_r. That is why these color spaces are important in digital image processing, even though raw YIQ or YUV images are rarely encountered in practice.

YUV

YUV is the basis for the color encoding used in analog television in both the North American NTSC and the European PAL systems. The luminance component Y is computed, just as in Eqn. (8.6), from the RGB components as

$$Y = 0.299 \cdot R + 0.587 \cdot G + 0.114 \cdot B \tag{8.27}$$

under the assumption that the RGB values have already been gamma corrected according to the TV encoding standard ($\gamma_{\mathrm{NTSC}} = 2.2$ and $\gamma_{\mathrm{PAL}} = 2.8$, see Sec. 4.7) for playback. The UV components are computed from a weighted difference between the luminance and the blue or red components as

$$U = 0.492 \cdot (B - Y) \qquad \text{and} \qquad V = 0.877 \cdot (R - Y), \tag{8.28}$$

and the entire transformation from RGB to YUV is

$$\begin{pmatrix} Y \\ U \\ V \end{pmatrix} = \begin{pmatrix} 0.299 & 0.587 & 0.114 \\ -0.147 & -0.289 & 0.436 \\ 0.615 & -0.515 & -0.100 \end{pmatrix} \cdot \begin{pmatrix} R \\ G \\ B \end{pmatrix}. \tag{8.29}$$

The transformation from YUV back to RGB is found by inverting the matrix in Eqn. (8.29):

$$\begin{pmatrix} R \\ G \\ B \end{pmatrix} = \begin{pmatrix} 1.000 & 0.000 & 1.140 \\ 1.000 & -0.395 & -0.581 \\ 1.000 & 2.032 & 0.000 \end{pmatrix} \cdot \begin{pmatrix} Y \\ U \\ V \end{pmatrix}. \tag{8.30}$$

The color distributions in YUV-space for a set of natural images are shown in Fig. 8.18.

YIQ

The original NTSC system used a variant of YUV called YIQ (I for "in-phase", Q for "quadrature"), where both the U and V color vectors were rotated and mirrored such that

$$\begin{pmatrix} I \\ Q \end{pmatrix} = \begin{pmatrix} 0 & 1 \\ 1 & 0 \end{pmatrix} \cdot \begin{pmatrix} \cos\beta & \sin\beta \\ -\sin\beta & \cos\beta \end{pmatrix} \cdot \begin{pmatrix} U \\ V \end{pmatrix}, \tag{8.31}$$

where $\beta = 0.576$ (33°). The Y component is the same as in YUV. Although the YIQ has certain advantages with respect to bandwidth requirements it has been completely replaced by YUV [22, p. 240].

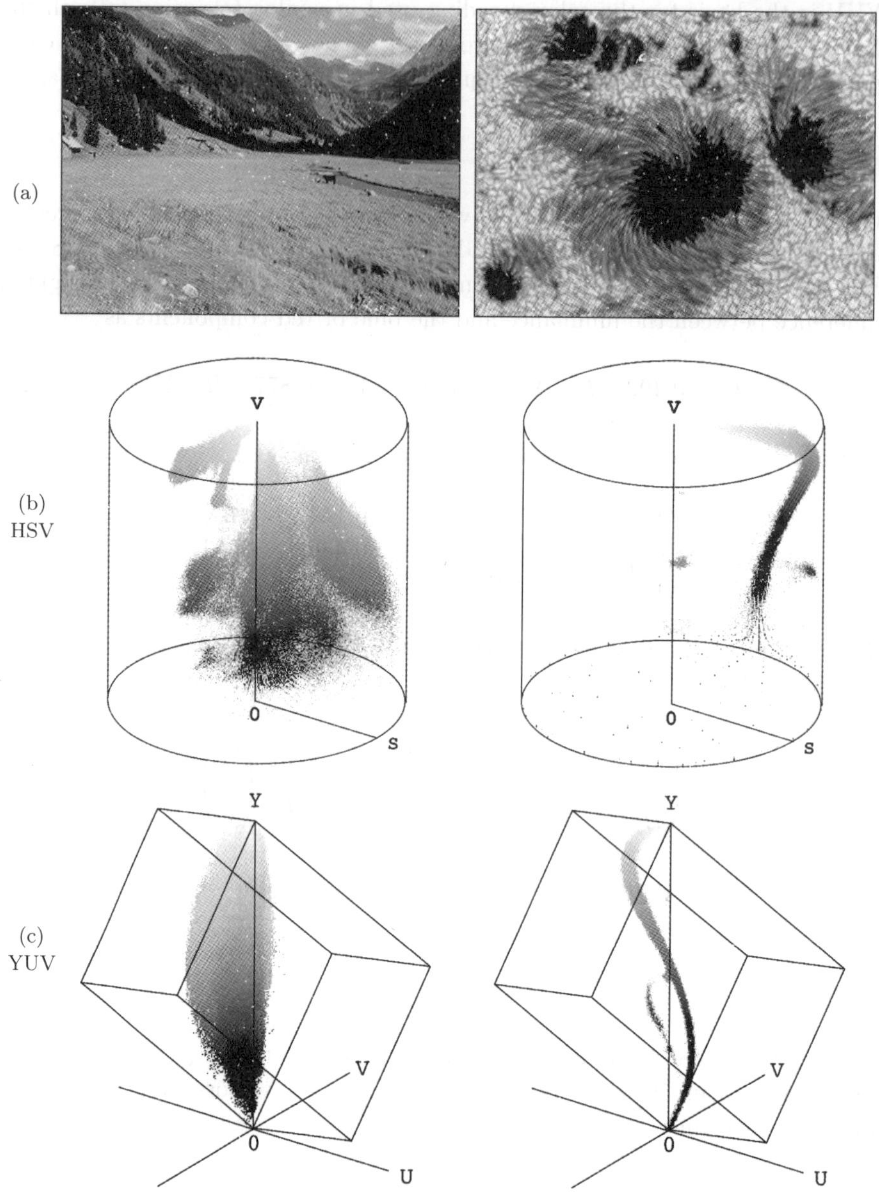

Figure 8.18 Examples of the color distribution of natural images in different color spaces. Original images (a); color distribution in HSV- (b), and YUV-space (c). See Fig. 8.9 for the corresponding distributions in RGB color space.

YC_bC_r

The YC_bC_r color space is an internationally standardized variant of YUV that is used for both digital television and image compression (for example, in JPEG). The chroma components C_b, C_r are (similar to U, V) difference values between the luminance and the blue and red components, respectively. In contrast to YUV, the weights of the RGB components for the luminance Y depend explicitly on the coefficients used for the chroma values C_b and C_r [35, p. 16]. For arbitrary weights w_B, w_R, the transformation is defined as

$$\begin{aligned} Y &= w_R \cdot R + (1 - w_B - w_R) \cdot G + w_B \cdot B, \\ C_b &= \frac{0.5}{1 - w_B} \cdot (B - Y), \\ C_r &= \frac{0.5}{1 - w_R} \cdot (R - Y), \end{aligned} \tag{8.32}$$

and the inverse transformation from YC_bC_r to RGB is

$$\begin{aligned} R &= Y + \frac{1 - w_R}{0.5} \cdot C_r, \\ G &= Y - \frac{w_B \cdot (1 - w_B) \cdot C_b \; - \; w_R \cdot (1 - w_R) \cdot C_r}{0.5 \cdot (1 - w_B - w_R)}, \\ B &= Y + \frac{1 - w_B}{0.5} \cdot C_b. \end{aligned} \tag{8.33}$$

The ITU[9] recommendation BT.601 [21] specifies the values $w_R = 0.299$ and $w_B = 0.114$ ($w_G = 1 - w_B - w_R = 0.587$). Using these values, the transformation becomes

$$\begin{pmatrix} Y \\ C_b \\ C_r \end{pmatrix} = \begin{pmatrix} 0.299 & 0.587 & 0.114 \\ -0.169 & -0.331 & 0.500 \\ 0.500 & -0.419 & -0.081 \end{pmatrix} \cdot \begin{pmatrix} R \\ G \\ B \end{pmatrix}, \tag{8.34}$$

and the inverse transformation becomes

$$\begin{pmatrix} R \\ G \\ B \end{pmatrix} = \begin{pmatrix} 1.000 & 0.000 & 1.403 \\ 1.000 & -0.344 & -0.714 \\ 1.000 & 1.773 & 0.000 \end{pmatrix} \cdot \begin{pmatrix} Y \\ C_b \\ C_r \end{pmatrix}. \tag{8.35}$$

Different weights are recommended based on how the color space is used; for example, ITU-BT.709 [20] recommends $w_R = 0.2125$ and $w_B = 0.0721$ to be used in digital HDTV production. The values of U, V, I, Q, and C_b, C_r may be both positive or negative. To encode C_b, C_r values to digital numbers, a suitable offset is typically added to obtain positive-only values, e. g., $128 = 2^7$ in case of 8-bit components.

---

[9] International Telecommunication Union (www.itu.int).

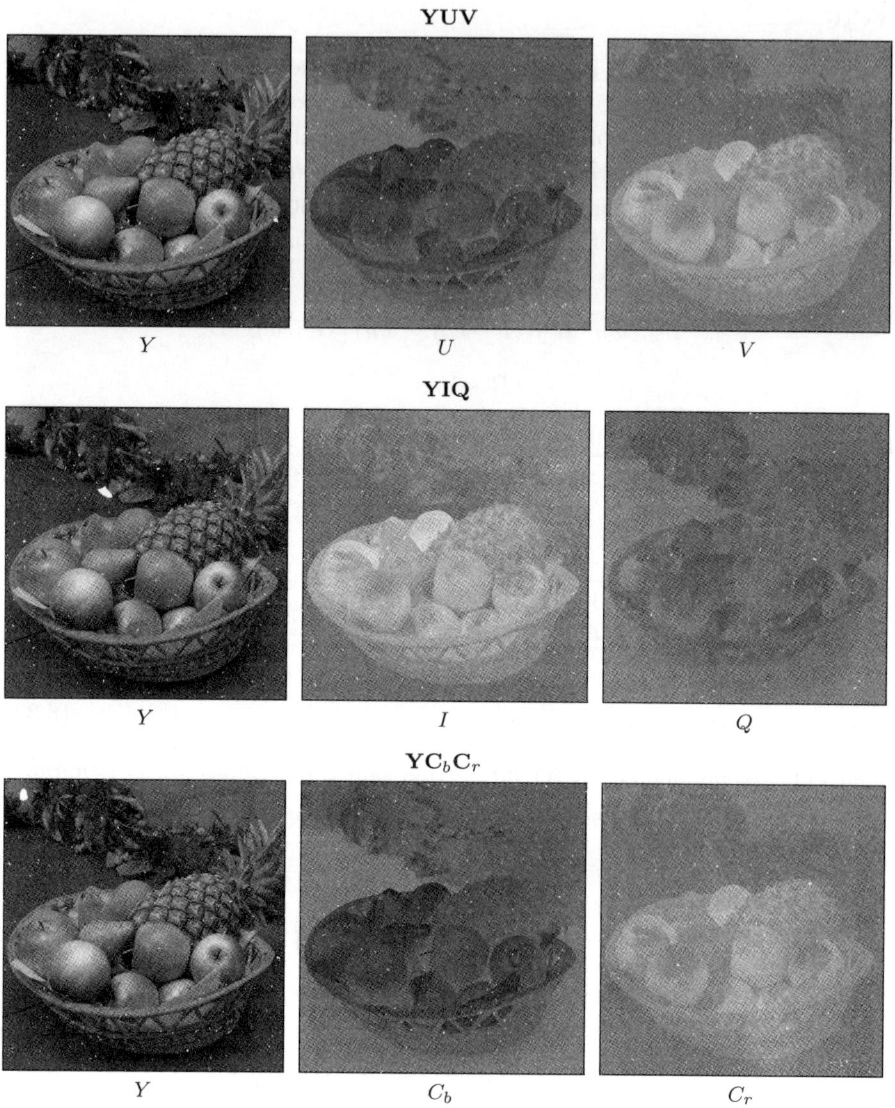

Figure 8.19 Comparing YUV-, YIQ- and YC$_b$C$_r$values. The Y values are identical in all three color spaces.

Figure 8.19 shows the three color spaces YUV, YIQ, and YC$_b$C$_r$ together for comparison. The U, V, I, Q, and C_b, C_r values in the right two frames have been offset by 128 so that the negative values are visible. Thus a value of zero is represented as medium gray in these images. The YC$_b$C$_r$ encoding is practically indistinguishable from YUV in these images since they both use very similar weights for the color components.

8.2.5 Color Spaces for Printing—CMY and CMYK

In contrast to the *additive* RGB color scheme (and its various color models), color printing makes use of a *subtractive* color scheme, where each printed color reduces the intensity of the reflected light at that location. Color printing requires a minimum of three primary colors; traditionally *cyan* (C), *magenta* (M) and *yellow* (Y)[10] have been used.

Using subtractive color mixing on a white background, $C = M = Y = 0$ (no ink) results in the color *white* and $C = M = Y = 1$ (complete saturation of all three inks) in the color *black*. A cyan-colored ink will absorb *red* (R) most strongly, magenta absorbs *green* (G), and yellow absorbs *blue* (B). The simplest form of the CMY model is defined as

$$\begin{aligned} C &= 1 - R, \\ M &= 1 - G, \\ Y &= 1 - B. \end{aligned} \tag{8.36}$$

In practice, the color produced by fully saturating all three inks is not physically a true black. Therefore, the three primary colors C, M, Y are usually supplemented with a black ink (K) to increase the color range and coverage (gamut). In the simplest case, the amount of black is

$$K = \min(C, M, Y). \tag{8.37}$$

With rising levels of black, however, the intensity of the C, M, Y components can be gradually reduced. Many methods for reducing the primary dyes have been proposed and we look at three of them in the following.

CMY→CMYK (Version 1): In this simple variant the C, M, Y values are reduced linearly with increasing K and the modified components C', M', Y', K' are defined as

$$\begin{pmatrix} C' \\ M' \\ Y' \\ K' \end{pmatrix} = \begin{pmatrix} C-K \\ M-K \\ Y-K \\ K \end{pmatrix}. \tag{8.38}$$

CMY→CMYK (Version 2): The second variant corrects the color by reducing the C, M, Y components by $s = \frac{1}{1-K}$, resulting in stronger colors in the

[10] Note that in this case Y stands for *yellow* and has nothing to do with the Y luminance component in YUV or YC_bC_r.

dark areas of the image:

$$\begin{pmatrix} C' \\ M' \\ Y' \\ K' \end{pmatrix} = \begin{pmatrix} (C-K) \cdot s \\ (M-K) \cdot s \\ (Y-K) \cdot s \\ K \end{pmatrix}, \quad \text{with } s = \begin{cases} \frac{1}{1-K} & \text{for } K < 1 \\ 1 & \text{otherwise.} \end{cases} \tag{8.39}$$

In both versions, the K-component (as defined in Eqn. (8.37)) is used directly without modification, and all gray tones (that is, when $R = G = B$) are printed using black ink K', without any contribution from C', M', or Y'.

While both of these simple definitions are widely used, neither one produces high quality results. Figure 8.20 (a) compares the result from version 2 with that produced with Adobe Photoshop (Fig. 8.20 (c)). The difference in the cyan component C is particularly noticeable and also the amount of black (K) brighter areas of the image.

In practice, the required amounts of black K and C, M, Y depend so strongly on the printing process and the type of paper used that print jobs are routinely calibrated individually.

CMY→CMYK (Version 3): In print production, special transfer functions are applied to tune the results. For example, the Adobe PostScript interpreter [26, p. 345] specifies an *undercolor-removal function* $f_{\mathrm{UCR}}(K)$ for gradually reducing the CMY components and a separate *black-generation function* $f_{\mathrm{BG}}(K)$ for controlling the amount of black. These functions are used in the form

$$\begin{pmatrix} C' \\ M' \\ Y' \\ K' \end{pmatrix} = \begin{pmatrix} C - f_{\mathrm{UCR}}(K) \\ M - f_{\mathrm{UCR}}(K) \\ Y - f_{\mathrm{UCR}}(K) \\ f_{\mathrm{BG}}(K) \end{pmatrix}, \tag{8.40}$$

where $K = \min(C, M, Y)$ again (as defined in Eqn. (8.37)). The functions f_{UCR} and f_{BG} are usually nonlinear, and the resulting values C', M', Y', K' are scaled (typically by means of *clamping*) to the interval $[0, 1]$. The example shown in Fig. 8.20 (b) was produced using the functions

$$f_{\mathrm{UCR}}(K) = s_K \cdot K, \tag{8.41}$$

$$f_{\mathrm{BG}}(K) = \begin{cases} 0 & \text{for } K < K_0 \\ K_{\max} \cdot \frac{K-K_0}{1-K_0} & \text{for } K \geq K_0, \end{cases} \tag{8.42}$$

where $s_K = 0.1$, $K_0 = 0.3$, and $K_{\max} = 0.9$ (see Fig. 8.21). With this definition, f_{UCR} reduces the CMY components by 10% of the K value (by Eqn. (8.40)), which mostly affects the dark areas of the image with high K values. The effect of the function f_{BG} (Eqn. (8.42)) is that for values of $K < K_0$ (that is in the

Figure 8.20 RGB→CMYK conversion comparison. Simple conversion using Eqn. (8.39) (a), applying the *undercolor-removal* and *black-generation* functions of Eqn. (8.40) (b), and results obtained with Adobe Photoshop (c). The color intensities are shown inverted, i. e., darker areas represent higher CMYK color values. The simple conversion (a), in comparison with Photoshop's result (c), shows strong deviations in all color components, C and K in particular. The results in (b) are close to Photoshop's and could be further improved by tuning the corresponding function parameters.

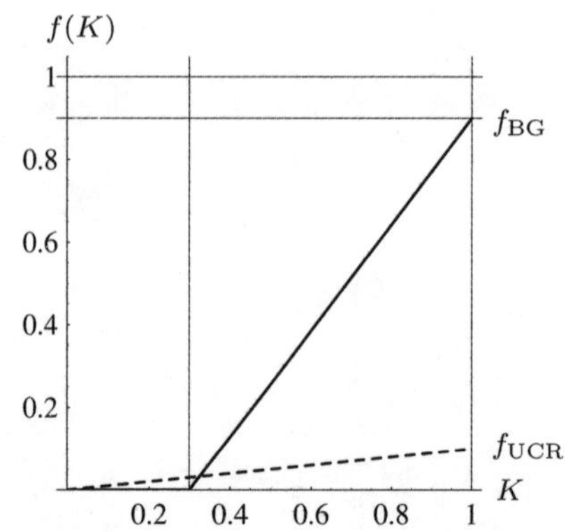

Figure 8.21 Examples of *undercolor-removal function* f_{UCR} (Eqn. (8.41)) and *black generation function* f_{BG} (Eqn. (8.42)). The parameter settings are $s_K = 0.1$, $K_0 = 0.3$, and $K_{max} = 0.9$.

light areas of the image), no black ink is added at all. In the interval $K = K_0 \ldots 1.0$, the black component is increased linearly up to the maximum value K_{max}. The result in Fig. 8.20 (b) is relatively close to the CMYK component values produced by Photoshop[11] in Fig. 8.20 (c). It could be further improved by adjusting the function parameters s_K, K_0, and K_{max} (Eqn. (8.40)).

Even though the results of this last variant (3) for converting RGB to CMYK are better, it is only a gross approximation and still too imprecise for professional work. As we discuss in Vol. 2 [6, Sec. 6], technically correct color conversions need to be based on precise, "colorimetric" grounds.

8.3 Statistics of Color Images

8.3.1 How Many Colors Are in an Image?

A minor but frequent task in the context of color images is to determine how many different colors are contained in a given image. One way of doing this would be to create and fill a histogram array with one integer element for each color and subsequently count all histogram cells with values greater than zero. But since a 24-bit RGB color image potentially contains $2^{24} = 16,777,216$ colors, the resulting histogram array (with a size of 64 megabytes) would be larger than the image itself in most cases!

A simple solution to this problem is to *sort* the pixel values in the (one-dimensional) pixel array such that all identical colors are placed next to each

[11] Actually Adobe Photoshop does not convert directly from RGB to CMYK. Instead, it first converts to, and then from, the CIE L*a*b* color space (see Vol. 2 [6, Sec. 6.2]).

```
1 import ij.process.ColorProcessor;
2 import java.util.Arrays;
3
4 public class ColorStatistics {
5
6   static int countColors (ColorProcessor cp) {
7     // duplicate pixel array and sort
8     int[] pixels = ((int[]) cp.getPixels()).clone();
9     Arrays.sort(pixels);
10
11    int k = 1; // image contains at least one color
12    for (int i = 0; i < pixels.length-1; i++) {
13      if (pixels[i] != pixels[i+1])
14        k = k + 1;
15    }
16    return k;
17  }
18
19 } // end of class ColorStatistics
```

Program 8.10 Counting the colors contained in an RGB image. The method `countColors()` first creates a copy of the one-dimensional RGB (`int`) pixel array (line 8), then sorts that array, and finally counts the transitions between contiguous blocks of identical colors.

other. The sorting order is of course completely irrelevant, and the number of contiguous color blocks in the sorted pixel vector corresponds to the number of different colors in the image. This number can be obtained by simply counting the transitions between neighboring color blocks, as shown in Prog. 8.10. Of course, we do not want to sort the original pixel array (which would destroy the image) but a copy of it, which can be obtained with Java's `clone()` method.[12] Sorting of the one-dimensional array in Prog. 8.10 is accomplished (in line 9) with the generic Java method `Arrays.sort()`, which is implemented very efficiently.

8.3.2 Color Histograms

We briefly touched on histograms of color images in Sec. 3.5, where we only considered the one-dimensional distributions of the image intensity and the individual color channels. For instance, the built-in ImageJ method `getHistogram()`, when applied to an object of type `ColorProcessor`, simply computes the intensity histogram of the corresponding gray values:

```
ColorProcessor cp;
int[] H = cp.getHistogram();
```

[12] Java arrays implement the methods of the root class `Object`, including the `clone()` method specified by the `Cloneable` interface (see also Appendix B.2.5).

As an alternative, one could compute the individual intensity histograms of the three color channels, although (as discussed in Sec. 3.5.2) these do not provide any information about the actual colors in this image. Similarly, of course, one could compute the distributions of the individual components of any other color space, such as HSV or L*a*b*.

A *full* histogram of an RGB image is three-dimensional and, as noted earlier, consists of $256 \times 256 \times 256 = 2^{24}$ cells of type `int` (for 8-bit color components). Such a histogram is not only very large[13] but also difficult to visualize.

2D color histograms

A useful alternative to the full 3D RGB histogram are two-dimensional histogram projections (Fig. 8.22). Depending on the axis of projection, we obtain 2D histograms with coordinates red-green (H_{RG}), red-blue (H_{RB}), or green-blue (H_{GB}), respectively, with the values

$$\begin{aligned} H_{\mathrm{RG}}(r,g) &\leftarrow \text{number of pixels with } I_{\mathrm{RGB}}(u,v)=(r,g,*), \\ H_{\mathrm{RB}}(r,b) &\leftarrow \text{number of pixels with } I_{\mathrm{RGB}}(u,v)=(r,*,b), \\ H_{\mathrm{GB}}(g,b) &\leftarrow \text{number of pixels with } I_{\mathrm{RGB}}(u,v)=(*,g,b), \end{aligned} \tag{8.43}$$

where $*$ denotes an arbitrary component value. The result is, independent of the original image size, a set of two-dimensional histograms of size 256×256 (for 8-bit RGB components), which can easily be visualized as images. Note that it is not necessary to obtain the full RGB histogram in order to compute the combined 2D histograms (see Prog. 8.11).

As the examples in Fig. 8.23 show, the combined color histograms do, to a certain extent, express the color characteristics of an image. They are therefore useful, for example, to identify the coarse type of the depicted scene or to estimate the similarity between images (see also Exercise 8.6).

8.4 Exercises

Exercise 8.1

Create an ImageJ plugin that rotates the individual components of an RGB color image; i. e., $R \rightarrow G \rightarrow B \rightarrow R$.

Exercise 8.2

Create an ImageJ plugin that shows the color table of an 8-bit indexed image as a new image with 16×16 rectangular color fields. Mark all unused color table entries in a suitable way. Look at Prog. 8.3 as a starting point.

[13] It may seem a paradox that, although the RGB histogram is usually much larger than the image itself, the histogram is not sufficient in general to reconstruct the original image.

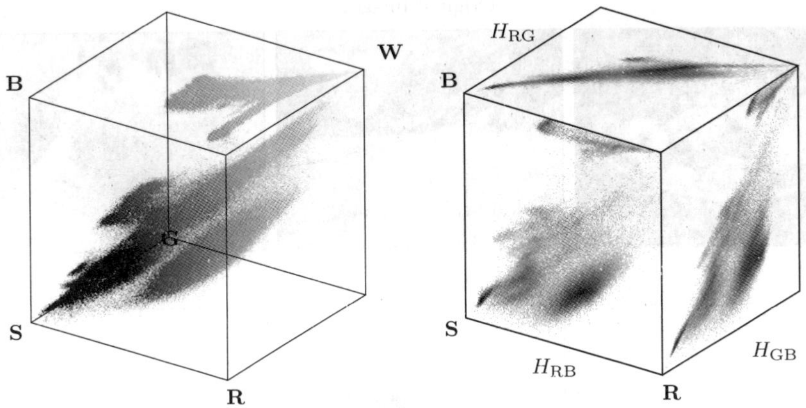

(a) (b)

Figure 8.22 Two-dimensional RGB histogram projections. Three-dimensional RGB cube illustrating an image's color distribution (a). The color points indicate the corresponding pixel colors and not the color frequency. The combined histograms for red-green (H_{RG}), red-blue (H_{RB}), and green-blue (H_{GB}) are 2D projections of the 3D histogram. The corresponding image is shown in Fig. 8.9 (a).

```
static int[][] get2dHistogram
            (ColorProcessor cp, int c1, int c2) {
  // c1, c2: R = 0, G = 1, B = 2
  int[] RGB = new int[3];
  int[][] H = new int[256][256]; // histogram array H[c1][c2]

  for (int v = 0; v < cp.getHeight(); v++) {
    for (int u = 0; u < cp.getWidth(); u++) {
      cp.getPixel(u, v, RGB);
      int i = RGB[c1];
      int j = RGB[c2];
      // increment corresponding histogram cell
      H[j][i]++; // i runs horizontal, j runs vertical
    }
  }
  return H;
}
```

Program 8.11 Method `get2dHistogram()` for computing a combined 2D color histogram. The color components (histogram axes) are specified by the parameters `c1` and `c2`. The color distribution `H` is returned as a two-dimensional `int` array. The method is defined in class `ColorStatistics` (Prog. 8.10).

Exercise 8.3

Show that a "desaturated" RGB pixel produced in the form $(r, g, b) \rightarrow (y, y, y)$, where y is the equivalent luminance value (see Eqn. (8.8)), has the luminance y as well.

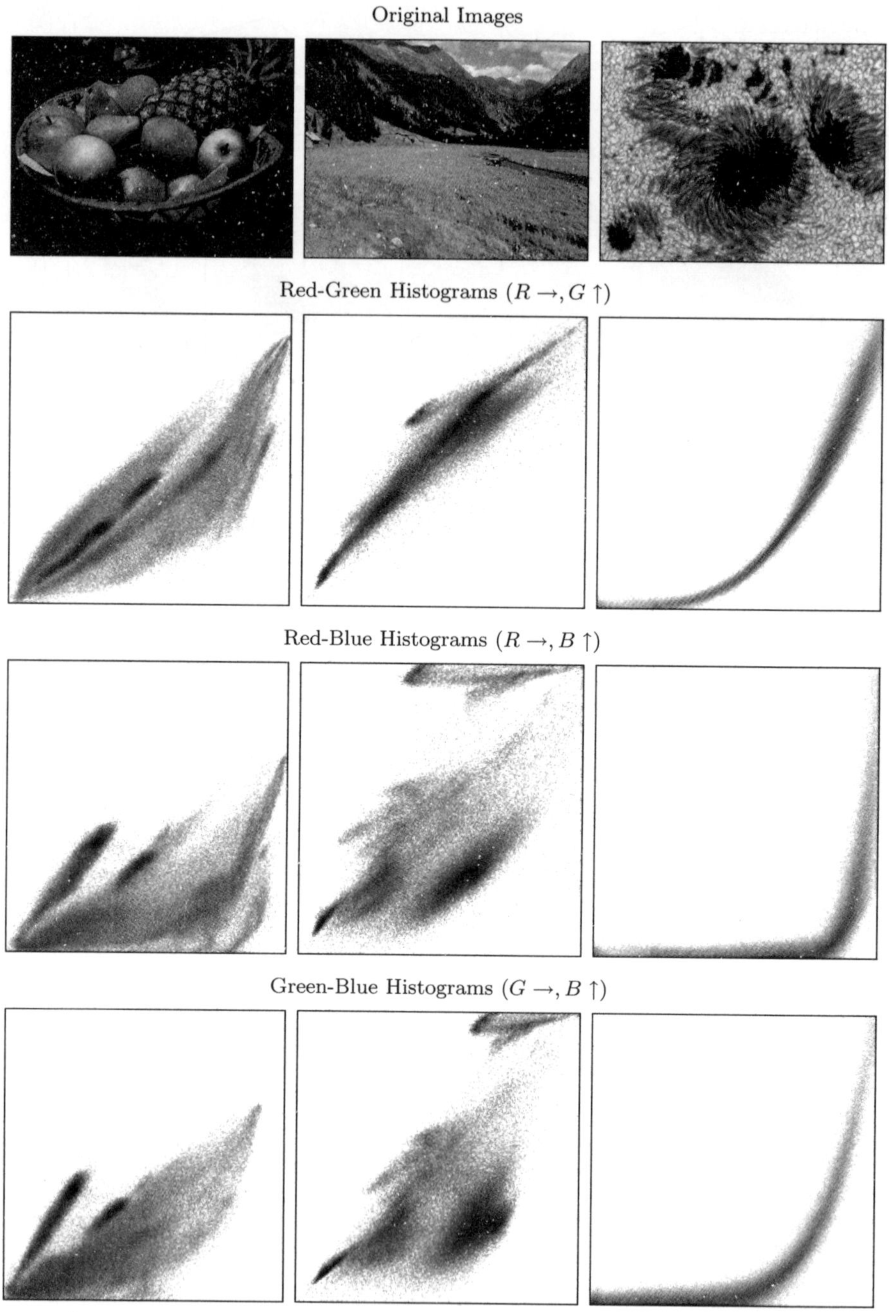

Figure 8.23 Combined color histogram examples. For better viewing, the images are inverted (dark regions indicate high frequencies) and the gray value corresponds to the logarithm of the histogram entries (scaled to the maximum entries).

Exercise 8.4

Extend the ImageJ plugin for desaturating color images in Prog. 8.5 such that the image is only modified inside the user-selected region of interest (ROI).

Exercise 8.5

Pseudocolors are sometimes used for displaying grayscale images (i. e., for viewing medical images with high dynamic range). Create an ImageJ plugin for converting 8-bit grayscale images to an indexed image with 256 colors, simulating the hues of glowing iron (from dark red to yellow and white).

Exercise 8.6

Determining the similarity between images of different sizes is a frequent problem (e. g., in the context of image data bases). Color statistics are commonly used for this purpose because they facilitate a coarse classification of images, such as landscape images, portraits, etc. However, two-dimensional color histograms (as described in Sec. 8.3.2) are usually too large and thus cumbersome to use for this purpose. A simple idea could be to split the 2D histograms or even the full RGB histogram into K regions (*bins*) and to combine the corresponding entries into a K-dimensional feature vector, which could be used for a coarse comparison. Develop a concept for such a procedure, and also discuss the possible problems.

A
Mathematical Notation

A.1 Symbols

The following symbols are used in the main text primarily with the denotations given below. While some symbols may be used for purposes other than the ones listed, the meaning should always be clear in the particular context.

$\{a, b, c, d, \ldots\}$ — A *set*; i. e., an unordered collection of distinct elements. A particular element x can be contained in a set at most once. A set may also be empty (denoted by $\{\ \}$).

$(a_1, a_2, \ldots a_n)$ — A *vector*; i. e., a fixed-size, ordered collection of elements of the same type. $(a_1, a_2, \ldots a_n)^T$ denotes the *transposed* (i. e., column) vector. In programming, vectors are usually implemented as one-dimensional arrays, with elements being referred to by position (index).

$[c_1, c_2, \ldots c_m]$ — A *sequence* or *list*; i. e., an ordered collection of elements of variable length. Elements can be added to the sequence (inserted) or deleted from the sequence. A sequence may be empty (denoted by $[\ ]$). In programming, sequences are usually implemented with dynamic data structures, such as linked lists. Java's *Collections* framework (see also Appendix B.2.7) provides numerous ready-to-use implementations.

$\langle \alpha_1, \alpha_2, \ldots \alpha_k \rangle$	A *tuple*; i. e., an ordered list of elements, each possibly of a different type. Tuples are typically implemented as *objects* (in Java or C++) or *structures* (in C) with elements being referred to by name.
$*$	Linear convolution operator (Sec. 5.3.1).
$\oplus$	Morphological dilation operator (Sec. 7.2.3).
$\ominus$	Morphological erosion operator (Sec. 7.2.4).
∂	Partial derivative operator (Sec. 6.2.1). For example, $\frac{\partial f}{\partial x}(x, y)$ denotes the *first* derivative of the function $f(x, y)$ along the x variable at position (x, y), $\frac{\partial^2 f}{\partial^2 x}(x, y)$ is the *second* derivative, etc.
∇	Gradient. ∇f is the vector of partial derivatives of a multidimensional function f (Sec. 6.2.1).
$\lfloor x \rfloor$	"Floor" of x, the largest integer $z \in \mathbb{Z}$ smaller than $x \in \mathbb{R}$ (i. e., $z = \lfloor x \rfloor \leq x$). For example, $\lfloor 3.141 \rfloor = 3$, $\lfloor -1.2 \rfloor = -2$.
a	Pixel value (usually $0 \leq a < K$).
$\mathrm{Arctan}(x, y)$	Inverse tangent function, similar to $\arctan(\frac{y}{x}) = \tan^{-1}(\frac{y}{x})$ but with two arguments and returning angles in the range $[-\pi, +\pi]$ (i. e., covering all four quadrants). It corresponds to the Java method `Math.atan2(y,x)` (Secs. 6.3, B.1.6).
$\mathrm{card}\{\ldots\}$	Cardinality (size) of a set, $\mathrm{card}\, A \equiv \lvert A \rvert$ (Sec. 3.1).
$\mathsf{h}(i)$	Histogram of an image at pixel value (or bin) i (Sec. 3.1).
$\mathsf{H}(i)$	Cumulative histogram of an image at pixel value (or bin) i (Sec. 3.6).
$I(u, v)$	Intensity or color value of the image I at (integer) position (u, v).
K	Number of possible pixel values.
M, N	Number of columns (width) and rows (height) of an image $(0 \leq u < M,\ 0 \leq v < N)$.
mod	Modulus operator: $(a \bmod b)$ is the remainder of the integer division a/b (Sec. B.1.2).

$\mathsf{p}(i)$	Probability density function (Sec. 4.6.1).
$\mathsf{P}(i)$	Probability distribution function or cumulative probability density (Sec. 4.6.1).
$\mathrm{round}(x)$	Rounding function: rounds x to the nearest integer. $\mathrm{round}(x) = \lfloor x + 0.5 \rfloor$.
$\mathrm{truncate}(x)$	Truncation function: truncates x toward zero to the closest integer. For example, $\mathrm{truncate}(3.141) = 3$, $\mathrm{truncate}(-2.5) = -2$.

A.2 Set Operators

$\lvert A \rvert$	The size (number of elements) of the set A (equivalent to $\mathrm{card}\, A$).
$\forall_x \ldots$	"All" quantifier (for all x, ...).
$\exists_x \ldots$	"Exists" quantifier (there is some x for which ...).
$\cup$	Set union (e. g., $A \cup B$).
$\cap$	Set intersection (e. g., $A \cap B$).
$\bigcup_{\mathcal{R}_i}$	Union over multiple sets $\mathcal{R}_i$.
$\bigcap_{\mathcal{R}_i}$	Intersection over multiple sets $\mathcal{R}_i$.

A.3 Algorithmic Complexity and $\mathcal{O}$ Notation

The term "complexity" describes the effort (i. e., computing time or storage) required by an algorithm or procedure to solve a particular problem in relation to the "problem size" n. Often complexity is reported in the literature using "big O" ($\mathcal{O}$) notation [18, Sec. 9.2], as in the following example. Consider a spreadsheet with 20 columns and 30 rows. Obviously, adding up all the entries in the spreadsheet requires performing $30 \cdot 20$ additions. We can be more general by representing the number of columns and rows by M and N, respectively, and saying it requires $M \cdot N$ additions. What if we want to replace each location with the sum of its eight neighbors? Then it would require $M \cdot N \cdot 8$ operations. If we compare these two algorithms, we see that, at their core, both require doing some number of operations $M \cdot N$ times. Since big O notation factors out constants (such as 8), we could say that the complexity of both of these algorithms is $\mathcal{O}(MN)$.

$\mathcal{O}(MN)$ is an upper bound on the number of operations an algorithm requires on an input of size MN. We can simplify this, since typical images have roughly the same number of rows and columns, by selecting the larger of the rows and columns $n = \max(M, N)$ and replacing it with n. Now, since we know $n \cdot n \geq M \cdot N$ we can say their complexity is $\mathcal{O}(n \cdot n)$ or, more commonly, $\mathcal{O}(n^2)$. Big O notation lets us compare *classes* of algorithms—in this case we discovered that both our algorithms belong to the $\mathcal{O}(n^2)$ class. This tells us that, no matter how much we optimize our code, at the heart our algorithm will require n^2 operations.

Similarly, the direct computation of the linear convolution (Sec. 5.3.1) for an image of size $n \times n$ and a convolution kernel of size $k \times k$ has the time complexity $\mathcal{O}(n^2 k^2)$. As another example, the *fast Fourier transform* (FFT, see Vol. 2 [6, Sec. 7.4.2]) of a signal vector of length $n = 2^k$ requires only $\mathcal{O}\big(n \log_2(n)\big)$ time.

Additional details on complexity can be found in any good book on computer algorithms, such as [1, 9].

B
Java Notes

As an undergraduate text for engineering curricula, this book assumes basic programming skills in a procedural language, such as C or Java. The examples in the main text should be easy to understand with the help of some introductory book on Java or one of the many online tutorials. Experience shows, however, that difficulties with some basic Java concepts pertain even at higher levels and frequently cause complications. The following sections aim at resolving some of these typical problem spots.

B.1 Arithmetic

Java is a "strongly typed" programming language, which means in particular that any variable has a fixed type that cannot be altered dynamically. Also, the result of an expression is determined by the types of the involved operands and *not* (in the case of an assignment) by the type of the "receiving" variable.

B.1.1 Integer Division

Division involving integer operands is a frequent cause of errors. If the variables `a` and `b` are both of type `int`, then the expression `(a / b)` is evaluated according to the rules of integer division. The result—the number of times `b` is contained in `a`—is again of type `int`. For example, after the Java statements

```
int a = 2;
int b = 5;
double c = a/b;
```

the value of c is *not* 0.4 but 0.0 because the expression a/b on the right produces the int value 0, which is then automatically converted to the double value 0.0.

If we wanted to evaluate a/b as a *floating-point* operation (as most pocket calculators do), at least one of the involved operands must be converted to a floating-point value, for example by an explicit type cast (double):

```
double c = (double) a / b;
```

Notice that the type cast (double) only applies to the immediately following term (a) and not the entire expression a / b; i. e., the value of the second operand (b) in this division is still of type int.

Example

Assume, for example, that we want to scale any pixel value a of an image such that the maximum pixel value $a_{\max}$ is mapped to 255 (see Ch. 4). In mathematical notation, the scaling of the pixel values is simply expressed as

$$c \leftarrow \frac{a}{a_{\max}} \cdot 255,$$

and it may be tempting to convert this 1:1 into Java code, such as

```
int a_max = ip.getMaxValue();
...
int a = ip.getPixel(u,v);
int c = (a / a_max) * 255;   ← PROBLEM!
ip.putPixel(u,v,a);
...
```

As we can easily predict, the resulting image will be all black (zero values), except those pixels whose value was a_max originally (they are set to 255). The reason is again the division (a / a_max) with two operands of type int, where the result is zero whenever the divisor (a_max) is greater than the dividend (a).

Of course, the entire operation could be performed in the floating-point domain by converting one of the operands (as shown earlier), but this is not even necessary in this case. Instead, we may simply swap the order of operations and start with the multiplication,

```
int c = a * 255 / a_max;
```

Why does this work? The subexpression a * 255 is evaluated first,[1] generating large intermediate values that pose no problem for the subsequent (integer) division. In addition, *rounding* should always be considered to obtain more accurate results when computing fractions of integers (see Sec. B.1.5).

[1] In Java, expressions at the same level are always evaluated in left-to-right order, and therefore no parentheses are required in this example (though they would not do any harm either).

B.1.2 Modulus Operator

The result of the modulus operator

$$a \bmod b$$

(used in several places in the main text) is defined [18, p. 82] as the remainder of the integer division a/b,

$$a \bmod b \triangleq \begin{cases} a & \text{for } b = 0 \\ a - b \cdot \left\lfloor \frac{a}{b} \right\rfloor & \text{otherwise.} \end{cases} \tag{B.1}$$

Unfortunately, this type of mod operator (or an equivalent library method) is not available in the standard Java API. Java's native `%` (*remainder*) operator, defined as

$$a \,\%\, b \triangleq a - b \cdot \text{truncate}\left(\frac{a}{b}\right) \quad \text{for } b \neq 0, \tag{B.2}$$

is often used in this context, but produces the same results only for *positive* operands $a \geq 0$ and $b > 0$. For example,

$$\begin{array}{rcrcr} 13 & \bmod & 4 & \rightarrow & 1 \\ 13 & \bmod & -4 & \rightarrow & -3 \\ -13 & \bmod & 4 & \rightarrow & 3 \\ -13 & \bmod & -4 & \rightarrow & -1 \end{array} \qquad \begin{array}{rcrcr} 13 & \% & 4 & \rightarrow & 1 \\ 13 & \% & -4 & \rightarrow & 1 \\ -13 & \% & 4 & \rightarrow & -1 \\ -13 & \% & -4 & \rightarrow & -1 \end{array}$$

The following Java method implements the mod operation according to the definition in Eqn. (B.1):

```
static int Mod(int a, int b) {
  if (b == 0)
    return a;
  if (a * b >= 0)
    return a - b * (a / b);
  else
    return a - b * (a / b - 1);
}
```

B.1.3 Unsigned Bytes

Most grayscale and indexed images in Java and ImageJ are composed of pixels of type `byte`, and the same holds for the individual components of most color images. A single byte consists of eight bits and can thus represent $2^8 = 256$ different bit patterns or values, usually mapped to the numeric range $0 \ldots 255$. Unfortunately, Java (unlike C and C++) does *not* provide a suitable "unsigned" 8-bit data type. The primitive Java type `byte` is "signed", using one of its eight bits for the ± sign, and can represent values in the range $-128 \ldots 127$.

Java's byte data can still be used to represent the values 0 to 255, but conversions must take place to perform proper arithmetic computation. For example, after execution of the statements

```
int  a = 200;
byte b = (byte) a;
```

the variables a (32-bit int) and b (8-bit byte) contain the binary patterns

```
a = 00000000000000000000000011001000
                        b = 11001000
```

respectively. Interpreted as a (signed) byte value, with the leftmost bit[2] as the sign bit, the variable b has the decimal value -56. Thus, after the statement

```
int a1 = b;                 // a1 == -56
```

the value of the new int variable a1 is -56! To (ab-)use signed byte data as *unsigned* data, we can circumvent Java's standard conversion mechanism by disguising the content of b as a logic (i. e., nonarithmetic) *bit pattern*; e. g., by

```
int  a2 = (0xff & b);      // a2 == 200
```

where 0xff (in hexadecimal notation) is an int value with the binary bit pattern 00000000000000000000000011111111 and & is the bitwise AND operator. Now the variable a2 contains the right integer value (200) and we thus have a way to use Java's (signed) byte data type for storing *unsigned* values. Within ImageJ, access to pixel data is routinely implemented in this way, which is considerably faster than using the convenience methods getPixel() and putPixel().

B.1.4 Mathematical Functions (Class Math)

Java provides the standard mathematical functions as static methods in class Math, as listed in Table B.1. The Math class is part of the java.lang package and thus requires no explicit import to be used. Most Math methods accept arguments of type double and also return values of type double. As a simple example, a typical use of the cosine function $y = \cos(x)$ is

```
double x;
double y = Math.cos(x);
```

Similarly, the Math class defines some common numerical constants as static variables; e. g., the value of π could be obtained by

```
double x = Math.PI;
```

[2] Java uses the standard "2s-complement" representation, where a sign bit = 1 stands for a negative value.

Table B.1 Methods and constants defined by Java's `Math` class.

`double abs(double a)`	`double max(double a, double b)`
`int abs(int a)`	`float max(float a, float b)`
`float abs(float a)`	`int max(int a, int b)`
`long abs(long a)`	`long max(long a, long b)`
`double ceil(double a)`	`double min(double a, double b)`
`double floor(double a)`	`float min(float a, float b)`
`double rint(double a)`	`int min(int a, int b)`
`long round(double a)`	`long min(long a, long b)`
`int round(float a)`	`double random()`
`double toDegrees(double rad)`	`double toRadians(double deg)`
`double sin(double a)`	`double asin(double a)`
`double cos(double a)`	`double acos(double a)`
`double tan(double a)`	`double atan(double a)`
`double atan2(double y, double x)`	
`double log(double a)`	`double exp(double a)`
`double sqrt(double a)`	`double pow(double a, double b)`
`double E`	`double PI`

B.1.5 Rounding

Java's `Math` class (confusingly) offers three different methods for rounding floating-point values:

```
double rint (double x)
long   round (double x)
int    round (float x)
```

For example, a `double` value `x` can be rounded to `int` in one of the following ways:

```
double x; int k;
k = (int) Math.rint(x);
k = (int) Math.round(x);
k = Math.round((float)x);
```

If the operand `x` is known to be positive (as is typically the case with pixel values) rounding can be accomplished without using any method calls by

```
k = (int) (x + 0.5); // works for x ≥ 0 only!
```

In this case, the expression `(x + 0.5)` is first computed as a floating-point (`double`) value, which is then truncated (toward zero) by the explicit `(int)` typecast.

B.1.6 Inverse Tangent Function

The inverse tangent function $\varphi = \tan^{-1}(a)$ or $\varphi = \arctan(a)$ is used in several places in the main text. This function is implemented by the method `atan(double a)` in Java's `Math` class (Table B.1). The return value of `atan()` is in the range $[-\frac{\pi}{2} \ldots \frac{\pi}{2}]$ and thus restricted to only two of the four quadrants. Without any additional constraints, the resulting angle is ambiguous. In many practical situations, however, a is given as the ratio of two catheti $(\Delta x, \Delta y)$ of a right-angled triangle in the form

$$\varphi = \tan^{-1}\Bigl(\frac{\Delta y}{\Delta x}\Bigr),$$

for which we used the (self-defined) two-parameter function

$$\varphi = \mathrm{Arctan}(\Delta y, \Delta x)$$

in the main text. The function $\mathrm{Arctan}(\Delta y, \Delta x)$ is implemented by the static method `atan2(dy,dx)` in Java's `Math` class and returns an unambiguous angle φ in the range $[-\pi \ldots \pi]$; i. e., in any of the four quadrants of the unit circle.[3]

B.1.7 `Float` and `Double` (Classes)

The representation of floating-point numbers in Java follows the IEEE standard, and thus the types `float` and `double` include the values

`POSITIVE_INFINITY`
`NEGATIVE_INFINITY`
`NaN` ("not a number")

These values are defined as constants in the corresponding wrapper classes `Float` and `Double`, respectively. If such a value occurs in the course of some computation (e. g., `POSITIVE_INFINITY` as the result of dividing by zero),[4] Java continues without raising an error.

B.2 Arrays and Collections

B.2.1 Creating Arrays

Unlike in most traditional programming languages (such as FORTRAN or C), arrays in Java can be created *dynamically*, meaning that the size of an array can be specified at runtime using the value of some variable or arithmetic expression. For example:

[3] The function `atan2(dy,dx)` is available in most current programming languages, including Java, C, and C++.

[4] In Java, this only holds for floating-point operations. Integer division by zero still causes an *exception*.

```
int N = 20;
int[] A = new int[N];
int[] B = new int[N*N];
```

Once allocated, however, the size of any Java array is fixed and cannot be subsequently altered. For additional variability, Java provides a number of universal container classes (e. g., the class `Vector`) for a wide range of applications.

After its definition, an array variable can be assigned any other compatible array or the constant value `null`; e. g.,

```
A = B;      // A now points to B's data
B = null;
```

Through the assignment `A = B` above, the array initially referenced by `A` becomes unaccessible and thus turns into *garbage*. In contrast to C and C++, where unnecessary storage needs to be *de*allocated explicitly, this is taken care of in Java by its built-in "garbage collector". It is also convenient that newly created arrays of numerical element types (`int`, `float`, `double`, etc.) are automatically initialized to zero.

B.2.2 Array Size

Since an array may be created dynamically, it is important that its actual size can be determined at runtime. This is done by accessing the `length` attribute[5] of the array:

```
int k = A.length; // number of elements in A
```

It may be surprising that Java arrays may have *zero* (not `null`) elements! If an array has more than one dimension, the size (`length`) along every dimension must be derived separately. The size is a property of the array itself and can therefore be obtained inside any method from array arguments passed to it. Thus (unlike in C, for example) it is not necessary to pass the size of an array as a separate function argument.

B.2.3 Accessing Array Elements

In Java, the index of the first array element is always 0 and the index of the last element is $N-1$ for an array with a total of N elements. To iterate through a one-dimensional array `A` of arbitrary size, one would typically use a construct like

```
for (int i = 0; i < A.length; i++) {
    // do something with A[i]
}
```

[5] Notice that the `length` attribute of an array is not a method!

Since images in Java and ImageJ are stored as one-dimensional arrays (accessible through the `ImageProcessor` method `getPixels()`), most point operations can be efficiently implemented in this way.[6]

B.2.4 Two-Dimensional Arrays

Multidimensional arrays are a common cause of misunderstanding. In Java, all arrays are one-dimensional, and multidimensional arrays are implemented as one-dimensional arrays of subarrays (Fig. B.1). If, for example, the 3×3 matrix

$$\boldsymbol{A} = \begin{pmatrix} A_{0,0} & A_{0,1} & A_{0,2} \\ A_{1,0} & A_{1,1} & A_{1,2} \\ A_{2,0} & A_{2,1} & A_{2,2} \end{pmatrix} = \begin{pmatrix} 1 & 2 & 3 \\ 4 & 5 & 6 \\ 7 & 8 & 9 \end{pmatrix}, \tag{B.3}$$

with elements a_{ij} (i being the *row* and j being the *column* index) is represented as a two-dimensional floating-point array,

```
double[][] A = {{1,2,3},
                {4,5,6},
                {7,8,9}};
```

then `A` is really a *one*-dimensional array containing three items, each of which is again a one-dimensional array of type `double` (see Fig. B.1).

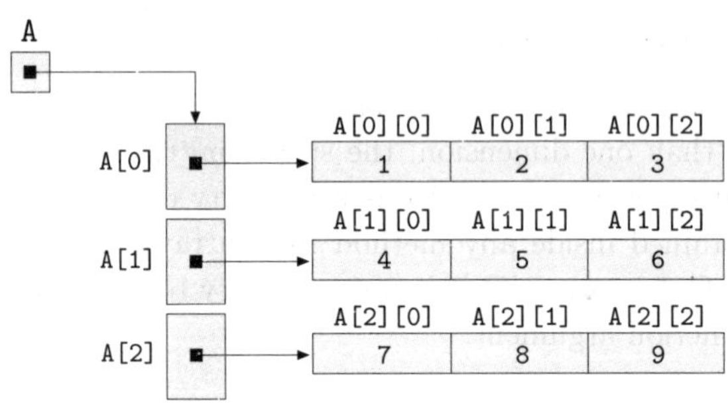

Figure B.1 Multidimensional arrays are implemented in Java as *one*-dimensional arrays whose elements are again one-dimensional arrays.

The usual assumption is that the array elements are arranged in *row-first* ordering, as illustrated in Fig. B.1. The first index thus corresponds to the row number *row* and the second index corresponds to the column number *col*,

$$a_{row,col} \equiv \texttt{A[row][col]} \qquad \text{or} \qquad a_{i,j} \equiv \texttt{A[i][j]}.$$

[6] See Prog. 7.1 in Sec. 7.6 of the ImageJ Short Reference [5] for an example.

So here the first array index runs downwards in the matrix and the second index runs to the right. This is quite convenient, because the array initialization in the code segment above looks exactly the same as the original matrix in Eqn. (B.3).

However, if the matrix represents an *image* or *filter kernel*, we usually associate the row index with the *vertical* coordinate v (or j) and the column index with the *horizontal* coordinate u (or i)—so the ordering of indices is reversed! For example, if we represent the filter kernel

$$H(i,j) = \begin{bmatrix} H(0,0) & H(1,0) & H(2,0) \\ H(0,1) & H(1,1) & H(2,1) \\ H(0,2) & H(1,2) & H(2,2) \end{bmatrix} = \begin{bmatrix} -1 & -2 & 0 \\ -2 & 0 & 2 \\ 0 & 2 & 1 \end{bmatrix}$$

(with i, j denoting the horizontal and vertical coordinate, respectively) as a two-dimensional Java array,

```
double[][] H = {{-1,-2, 0},
                {-2, 0, 2},
                { 0, 2, 1}};
```

then the indices must be *reversed* in order to access the right elements. In this particular case,

$$H(i,j) \equiv \texttt{H[j][i]}.$$

This scheme was used, for example, for implementing the 3×3 filter plugin in Prog. 5.2 (p. 105).

Size of Multi-Dimensional Arrays

The size of a multidimensional array can be obtained by querying the size of its subarrays. For example, given the following three-dimensional array with dimensions $P \times Q \times R$,

```
int[][][] B = new int[P][Q][R];
```

the size of `B` along its three dimensions is obtained by the statements

```
int p = B.length;          // = P
int q = B[0].length;       // = Q
int r = B[0][0].length;    // = R
```

At least this works for "rectangular" Java arrays, i. e., multidimensional arrays with all subarrays at the same level having *identical* length. If this is not the case, the length of each (one-dimensional) subarray must be determined individually to avoid "index-out-of-bounds" errors. Thus a "bullet-proof" iteration over all elements of a three-dimensional—potentially "non-rectangular"—array `C` could be implemented as follows:

```
import java.lang.reflect.Array;

public static Object duplicateArray(Object orig) {
  Class origClass = orig.getClass();
  if (!origClass.isArray())
    return null; // no array to duplicate
  Class compType = origClass.getComponentType();
  int n = Array.getLength(orig);
  Object dup = Array.newInstance(compType, n);
  if (compType.isArray()) // array elements are arrays again:
    for (int i = 0; i < n; i++)
      Array.set(dup, i, duplicateArray(Array.get(orig, i)));
  else // array elements are objects or primitives:
    System.arraycopy(orig, 0, dup, 0, n);
  return dup;
}
```

Program B.1 Utility method `duplicateArray()` for cloning arrays of any element type and dimensionality. Objects inside the array are not duplicated.

```
for (int i = 0; i < C.length; i++) {
  for (int j = 0; j < C[i].length; j++) {
    for (int k = 0; k < C[i][j].length; k++) {
      // do something with C[i][j][k]
    }
  }
}
```

B.2.5 Cloning Arrays

Java arrays implement the standard `java.lang.Cloneable` interface and provide `clone()` methods to perform a single-level ("shallow") form of duplication; i. e., to make a copy of the top-level structure of the array. Applied to a one-dimensional array of primitive element type, e. g.,

```
int[] A1 = {1,2,3,4};
int[] A2 = (int[]) A1.clone();
```

the result `A2` is an exact and independent copy of the array `A1`, as one would expect. If the original array contains real (i. e., nonprimitive) Java *objects*, `clone()` does *not* duplicate the individual objects themselves, but the cells of both arrays refer to the same original objects.

Similarly, applying `clone()` to a two-dimensional (or multidimensional) array duplicates only the top-level structure of that array but none of its subarrays. Java has no standard method for doing a *full-depth* duplication of multidimensional arrays. The (nontrivial) method `duplicateArray()` in Prog. B.1 shows how this could be accomplished recursively for arrays of any element type and dimensionality.

B.2.6 Arrays of Objects, Sorting

In Java, as mentioned earlier, we can create arrays dynamically; i. e., the size of an array can be specified during execution. This is convenient because we can adapt the size of the arrays to the actual problem. For example, we could write

```
Corner[] cornerArray = new Corner[n];
```

to create an array that can hold `n` objects of type `Corner` (as defined in Vol. 2 [6, Sec. 4.3]). But be aware that the new array is not filled with corners yet but initialized with `null` (i. e., empty references), so the array is really empty. We can insert a `Corner` object into its first (or any other) cell by

```
cornerArray[0] = new Corner(10,20,6789.0f);
```

Arrays can be sorted quickly using the static utility methods in the `java.util.Arrays` class,

`Arrays.sort(`*type*`[] arr)`

where `arr` can be any array of primitive *type* (`int`, `float`, etc.) or an array of objects. In the latter case, the array may not have `null` entries. Also, the class of every contained object must implement the `Comparable` interface, i. e., provide a public method

`int compareTo(Object obj)`

that must return an `int` value of -1, 0, or 1, depending upon the intended order relation to the other object `obj`. For example, within the `Corner` class, the `compareTo()` method could be defined as follows:

```
public int compareTo (Object obj){    // in class Corner
  Corner c2 = (Corner) obj;
  if (this.q > c2.q) return -1;
  if (this.q < c2.q) return 1;
  else return 0;
}
```

which implicitly assumes that objects of class `Corner` need never be compared with any other type of object.[7]

In summary, arrays are highly efficient data structures that allow fast searching and sorting and therefore should be used whenever fixed size is not a problem.

---

[7] Note that the typecast `(Corner)obj` (line 2 in method `compareTo`) is potentially dangerous and will create a runtime exception if `obj` is not of type `Corner`.

B.2.7 Collections

Once created, arrays in Java are of fixed size and cannot be expanded or shrunk. To use an array for collecting the corners detected in an image may thus not be a good idea because we do not know a priori how many corners the image contains. If we make the initial array too small, we will run out of space during the process. If we make the array as large as possibly needed, we will probably waste a lot of memory most of the time.

When we try to extract entities (e. g., corner points) from images, we do not know in advance how many of them we are going to find. Also, the properties of these items of interest may vary. This is a frequent situation, and while most simple processes in digital imaging are done with fixed-sized arrays of numbers, dynamic data structures are often needed for advanced tasks. Incidentally, this is also one of Java's strongest aspects. In fact, Java provides a complete collection framework with several convenient data structures that would be complicated to implement by oneself.

A "collection" represents a group of objects, known as its elements. So arrays, which we have been using over and over again, are of course collections. The Java collections framework is a unified architecture for representing and manipulating collections, allowing them to be manipulated independently of the details of their representation. It reduces programming effort while delivering high performance. It allows for interoperability among unrelated APIs, reduces effort in designing and learning new APIs, and fosters software reuse. The framework is based on six collection interfaces. It includes implementations of these interfaces and algorithms to manipulate them. Some types of collections allow duplicate elements and others do not, and some collections are ordered and others unordered.

The Java SDK does not provide any *direct* implementations of this interface but implements more specific subinterfaces such as `Set` and `List`. This interface is typically used to pass collections around and manipulate them where maximum generality is desired. Concrete implementations of the `Collection` interface include the classes `Vector` and `ArrayList`, as well as `HashSet` for the convenient construction of hash tables.

Additional details and application examples can be found in the Java SDK documentation[8] and the Java Collections tutorial.[9] For general hints on effective programming in Java, the classic book by Bloch [4] is a particularly valuable source.

[8] http://java.sun.com/javase/reference/

[9] http://java.sun.com/docs/books/tutorial/collections/

Bibliography

[1] A. V. Aho, J. E. Hopcroft, and J. D. Ullman. "The Design and Analysis of Computer Algorithms". Addison-Wesley, Reading, MA (1974).

[2] K. Arnold, J. Gosling, and D. Holmes. "The Java Programming Language". Addison-Wesley, Reading, MA, fourth ed. (2005).

[3] W. Bailer. "Writing ImageJ Plugins—A Tutorial" (2003). http://www.imagingbook.com.

[4] J. Bloch. "Effective Java Programming Language Guide". Addison-Wesley, Reading, MA (2001).

[5] W. Burger and M. J. Burge. "ImageJ Short Reference for Java Developers" (2008). http://www.imagingbook.com.

[6] W. Burger and M. J. Burge. "Principles of Image Processing—Core Algorithms". Springer, New York (2009).

[7] P. J. Burt and E. H. Adelson. The Laplacian pyramid as a compact image code. *IEEE Transactions on Communications* **31**(4), 532–540 (1983).

[8] J. F. Canny. A computational approach to edge detection. *IEEE Trans. on Pattern Analysis and Machine Intelligence* **8**(6), 679–698 (1986).

[9] T. H. Cormen, C. E. Leiserson, R. L. Rivest, and C. Stein. "Introduction to Algorithms". MIT Press, Cambridge, MA, second ed. (2001).

[10] L. S. Davis. A survey of edge detection techniques. *Computer Graphics and Image Processing* **4**, 248–270 (1975).

[11] B. Eckel. "Thinking in Java". Prentice Hall, Englewood Cliffs, NJ, fourth ed. (2006). Earlier versions available online.

[12] N. Efford. "Digital Image Processing—A Practical Introduction Using Java". Pearson Education, Upper Saddle River, NJ (2000).

[13] D. Flanagan. "Java in a Nutshell". O'Reilly, Sebastopol, CA, fifth ed. (2005).

[14] J. D. Foley, A. van Dam, S. K. Feiner, and J. F. Hughes. "Computer Graphics: Principles and Practice". Addison-Wesley, Reading, MA, second ed. (1996).

[15] A. Ford and A. Roberts. "Colour Space Conversions" (1998). http://www.poynton.com/PDFs/coloureq.pdf.

[16] A. S. Glassner. "Principles of Digital Image Synthesis". Morgan Kaufmann Publishers, San Francisco (1995).

[17] R. C. Gonzalez and R. E. Woods. "Digital Image Processing". Addison-Wesley, Reading, MA (1992).

[18] R. L. Graham, D. E. Knuth, and O. Patashnik. "Concrete Mathematics: A Foundation for Computer Science". Addison-Wesley, Reading, MA, second ed. (1994).

[19] R. W. G. Hunt. "The Reproduction of Colour". Wiley, New York, sixth ed. (2004).

[20] International Telecommunications Union, ITU, Geneva. "ITU-R Recommendation BT.709-3: Basic Parameter Values for the HDTV Standard for the Studio and for International Programme Exchange" (1998).

[21] International Telecommunications Union, ITU, Geneva. "ITU-R Recommendation BT.601-5: Studio Encoding Parameters of Digital Television for Standard 4:3 and Wide-Screen 16:9 Aspect Ratios" (1999).

[22] K. Jack. "Video Demystified—A Handbook for the Digital Engineer". LLH Publishing, Eagle Rock, VA, third ed. (2001).

[23] B. Jähne. "Practical Handbook on Image Processing for Scientific Applications". CRC Press, Boca Raton, FL (1997).

[24] B. Jähne. "Digitale Bildverarbeitung". Springer-Verlag, Berlin, fifth ed. (2002).

[25] A. K. Jain. "Fundamentals of Digital Image Processing". Prentice Hall, Englewood Cliffs, NJ (1989).

[26] J. KING. Engineering color at Adobe. In P. GREEN AND L. MACDONALD, editors, "Colour Engineering", ch. 15, pp. 341–369. Wiley, New York (2002).

[27] R. A. KIRSCH. Computer determination of the constituent structure of biological images. *Computers in Biomedical Research* **4**, 315–328 (1971).

[28] T. LINDEBERG. Feature detection with automatic scale selection. *International Journal of Computer Vision* **30**(2), 77–116 (1998).

[29] D. MARR AND E. HILDRETH. Theory of edge detection. *Proceedings of the Royal Society of London, Series B* **207**, 187–217 (1980).

[30] J. MIANO. "Compressed Image File Formats". ACM Press, Addison-Wesley, Reading, MA (1999).

[31] P. A. MLSNA AND J. J. RODRIGUEZ. Gradient and laplacian-type edge detection. In A. BOVIK, editor, "Handbook of Image and Video Processing", pp. 415–431. Academic Press, New York (2000).

[32] J. D. MURRAY AND W. VANRYPER. "Encyclopedia of Graphics File Formats". O'Reilly, Sebastopol, CA, second ed. (1996).

[33] T. PAVLIDIS. "Algorithms for Graphics and Image Processing". Computer Science Press / Springer-Verlag, New York (1982).

[34] W. S. RASBAND. "ImageJ". U.S. National Institutes of Health, MD (1997–2007). http://rsb.info.nih.gov/ij/.

[35] I. E. G. RICHARDSON. "H.264 and MPEG-4 Video Compression". Wiley, New York (2003).

[36] L. G. ROBERTS. Machine perception of three-dimensional solids. In J. T. TIPPET, editor, "Optical and Electro-Optical Information Processing", pp. 159–197. MIT Press, Cambridge, MA (1965).

[37] J. C. RUSS. "The Image Processing Handbook". CRC Press, Boca Raton, FL, third ed. (1998).

[38] Y. SCHWARZER, editor. "Die Farbenlehre Goethes". Westerweide Verlag, Witten (2004).

[39] N. SILVESTRINI AND E. P. FISCHER. "Farbsysteme in Kunst und Wissenschaft". DuMont, Cologne (1998).

[40] M. STOKES AND M. ANDERSON. "A Standard Default Color Space for the Internet—sRGB". Hewlett-Packard, Microsoft, www.w3.org/Graphics/Color/sRGB.html (1996).

[41] A. WATT. "3D Computer Graphics". Addison-Wesley, Reading, MA, third ed. (1999).

[42] A. WATT AND F. POLICARPO. "The Computer Image". Addison-Wesley, Reading, MA (1999).

[43] G. WOLBERG. "Digital Image Warping". IEEE Computer Society Press, Los Alamitos, CA (1990).

[44] T. Y. ZHANG AND C. Y. SUEN. A fast parallel algorithm for thinning digital patterns. *Communications of the ACM* **27**(3), 236–239 (1984).

Index

D

E

F

About the Authors

Wilhelm Burger received a Master's degree in Computer Science from the University of Utah (Salt Lake City) and a doctorate in Systems Science from Johannes Kepler University in Linz, Austria. As a post-graduate researcher at the Honeywell Systems & Research Center in Minneapolis and the University of California at Riverside, he worked mainly in the areas of visual motion analysis and autonomous navigation. In the Austrian research initiative on digital imaging, he was engaged in projects on generic object recognition and biometric identification. Since 1996, he has been the director of the Digital Media degree programs at the Upper Austria University of Applied Sciences at Hagenberg. Personally the author appreciates large-engine vehicles and (occasionally) a glass of dry "Veltliner".

Mark J. Burge received a BA degree from Ohio Wesleyan University, a MSc in Computer Science from the Ohio State University, and a doctorate from Johannes Kepler University in Linz, Austria. He spent several years as a researcher in Zürich, Switzerland at the Swiss Federal Institute of Technology (ETH), where he worked in computer vision and pattern recognition. As a post-graduate researcher at the Ohio State University, he was involved in the "Image Understanding and Interpretation Project" sponsored by the NASA Commercial Space Center. He earned tenure within the University System of Georgia as an associate professor in computer science and served as a Program Director at the National Science Foundation. Currently he is a Principal at Noblis (Mitretek) in Washington D.C. Personally, he is an expert on classic Italian espresso machines.

About this Book Series

The complete manuscript for this book was prepared by the authors "camera-ready" in LaTeX using Donald Knuth's Computer Modern fonts. The additional packages algorithmicx (by Szász János) for presenting algorithms, listings (by Carsten Heinz) for listing progam code, and psfrag (by Michael C. Grant and David Carlisle) for replacing text in graphics were particularly helpful in this task. Most illustrations were produced with Macromedia Freehand (now part of Adobe), function plots with Mathematica, and images with ImageJ or Adobe Photoshop. All book figures, test images in color and full resolution, as well as the Java source code for all examples are available at the book's support site: www.imagingbook.com.

About the Author